Calculus Problem Workbook
for
Hecht's
PHYSICS: CALCULUS

Zvonimir Hlousek
California State University Long Beach

and

Regina L. Neiman
Illinois Wesleyan University

Brooks/Cole Publishing Company

I(T)P® An International Thomson Publishing Company

Pacific Grove • Albany • Bonn • Boston • Cincinnati • Detroit
London • Madrid • Melbourne • Mexico City • New York • Paris
San Francisco • Singapore • Tokyo • Toronto • Washington

Assistant Editor: *Elizabeth Barelli Rammel*
Cover Design: *Vernon T. Boes*
Cover Photo: *Tom Skrivan*
Compositor: *Eigentype Compositors*
Director Marketing Communications: *Margaret Parks*
Editorial Associate: *Beth Wilbur*
Production Editor: *Mary Vezilich*
Printing and Binding: *Malloy Lithographing*

 The ITP logo is a registered trademark under license.

For more information, contact:

BROOKS/COLE PUBLISHING COMPANY
511 Forest Lodge Road
Pacific Grove, CA 93950
USA

International Thomson Editores
Campos Eliseos 385, Piso 7
Col. Polanco
11560 México D. F. México

International Thomson Publishing Europe
Berkshire House 168-173
High Holborn
London WC1V 7AA
England

International Thomson Publishing GmbH
Königswinterer Strasse 418
53227 Bonn
Germany

Thomas Nelson Australia
102 Dodds Street
South Melbourne, 3205
Victoria, Australia

International Thomson Publishing Asia
221 Henderson Road
#05-10 Henderson Building
Singapore 0315

Nelson Canada
1120 Birchmount Road
Scarborough, Ontario
Canada M1K 5G4

International Thomson Publishing Japan
Hirakawacho Kyowa Building, 3F
2-2-1 Hirakawacho
Chiyoda-ku, Tokyo 102
Japan

Printed in the United States of America

5   4   3   2   1

ISBN 0-534-32399-5

TABLE OF CONTENTS

When writing a student study guide, such as this one, it is important to be explicit about how the guide is to be used. With this in mind, we would like to explain the purpose for the two parts of this guide. The first part consists of examples with detailed solutions, designed to explain the method of solving problems. These examples are meant to stress the basics of the given chapter in order to help you better understand how to approach calculus based problems. The second part consists of exercises for you to do, with answers in the back of the book. We recommend that you attempt these problems and if you can't complete them then go back and study the material some more. If you need to, ask your teacher to help you. Remember, this guide will not teach you calculus – you had to have taken a calculus course or should be taking one concurrently with the physics course. This guide will however, help you apply calculus to physics problems.

These exercises were designed so that you can do them in any order you desire. And by no means, must you do all the problems in the study guide. A common misconception of students who are just beginning to study physics is that physics can be learned merely by reading. This is not correct, and is definitely not the purpose of this guide. This guide is meant as a workbook, a place for you to do problems.

The problems in this guide fall into two broad categories – easy problems and not so easy ones! Approximately 60% of the exercises belong to *the easy* category. You can use them to check how comfortable you are with the material. The remaining 40% are more challenging problems. As you may know, the calculus was conceived out of physics. Since its invention, calculus has been one of the basic tools of a physicist. It has enabled us to understand new phenomena occuring in nature. The logic of calculus is at the very roots of physics; without understanding it we cannot truly understand the subject matter of physics. The purpose of the more challenging problems in this workbook is to help you understand the subject at a deeper level.

As you are doing the problems in this guide, you might notice a difference in style as you go from chapter to chapter. We hope that this helps you as a student. We are two different authors and our approaches to the subject are different as a result of differing perspectives, intuitions and experiences. There is no single way to approach even the simplest problem in physics. We have both tried to present solutions in a way we think you would approach them. The textbook that this workbook accompanies is an excellent resource and you should consult it often.

We have tried hard to make this workbook error free. However, some errors have possibly been overlooked. If you have corrections or comments concerning this study guide, we would appreciate it if you write to us. For the odd numbered chapters, please contact Prof. Hlousek at California State University - Long Beach, Department of Physics, 1250 Bellflower Blvd., Long Beach, CA 90840; e-mail: hlousek@csulb.edu. For the even numbered chapters, please contact Prof. Neiman at Illinois Wesleyan University, Department of Physics, Bloomington, IL 61702; e-mail: rneiman@titan.iwu.edu.

This guide was printed from camera ready copy typeset in TEX. Illustrations were created by AutoCad and further rendered in Adobe Illustrator before they were embedded into TEX. We would like to thank several people who helped us put this guide together: Ms. Vesna Medak-Hlousek who created the art under our instructions. All of the art that looks good is to her credit—in the instances you find the art unsatisfactory blame us because we failed to present clear instructions; Ms. Elizabeth Rammel, our editor who kept us in place and made sure that the manuscript was written and ready; Dr. Eugene Hecht and Prof. Lee Young who marked our draft pages in red ink, so much that we could hardly recognize our own print; and Peter Vacek of Eigentype Compositors, Inc., for patiently listening and putting together the manuscript and art files into a final form.

Good luck! We hope you find this guide illuminating and useful.

Sincerely,

Zvonimir Hloušek and Regina Neiman

EXAMPLE 2.1

Tom the cat is chasing Jerry the mouse around the kitchen of their house. At one point, Jerry gets smart and decides to run through a small hole in the baseboards. Tom, without thinking, tries to follow Jerry, but he isn't small enough to fit through the hole. If Tom's position from the wall can be described as:

$$x(t) = (2.00\,\text{m/s})t - (1.50\,\text{m/s}^2)t^2,$$

calculate Tom's speed just before he hits the wall.

$x = 0.00$ mm

Figure: EG-2.1

SOLUTION

[Given: $x(t) = At - Bt^2$, where $A = 2.00\,\text{m/s}$ and $B = 1.50\,\text{m/s}^2$. Find: v when $x = 0$.] To get the speed of Tom, just before he hits the wall, we need, first, to get an expression for $v(t)$ and, second, we need to determine the time at which Tom hits the wall.

First, since $v = \frac{dx}{dt}$, we can get an expression for v:

$$v = \frac{dx}{dt} = \frac{d}{dt}(At - Bt^2)$$
$$v = A - 2Bt$$

Now, when does Tom hit the wall? We know Tom's position at the wall will be $x = 0$, so this will tell us what time Tom hits the wall: $0 = At - Bt^2$

$$At = Bt^2 \qquad \text{so} \qquad t = \frac{A}{B} = \frac{2.00\,\text{m/s}}{1.50\,\text{m/s}^2} = 1.3\overline{3}\,\text{s}.$$

Now, we can substitute for the time to get Tom's speed when he just hits the wall:

$$v = A - 2Bt$$
$$= (2.00\,\text{m/s}) - 2(1.50\,\text{m/s}^2)(1.3\overline{3}\,\text{s})$$
$$= -2.00\,\text{m/s}$$

So, Tom's speed is 2.00 m/s just before he hits the wall.

EXAMPLE 2.2

A bag of groceries is sliding around in the trunk of a car. The bag's position varies with time as:

$$\ell(t) = 0.500\,\text{m}\, e^{-(2.00\times 10^{-3}\,\text{s}^{-1})t} \cos(15.4\,\text{s}^{-1}\cdot t)$$

Calculate the speed of the bag as a function of time.

SOLUTION

[Given: $s(t) = Ae^{-Bt}\cos(Ct)$, where $A = 0.500\,\text{m}$, $B = 2.00 \times 10^{-3}\,\text{s}^{-1}$, and $c = 15.4\,\text{s}^{-1}$. Find: $v(t)$.] To calculate the velocity of the bag of groceries, we need

2

to differentiate the position with respect to time.

$$v = \frac{d}{dt}\ell(t) = \frac{d}{dt}\left[Ae^{-Bt}\cos(Ct)\right]$$
$$= -B(Ae^{-Bt}\cos Ct) - CAe^{-Bt}\sin(Ct)$$

[Remember to do the product rule to solve this derivative.] Now, we can plug in the values:

$$v(t) = -(2.00 \times 10^{-3}\,\mathrm{s}^{-1})(0.500\,\mathrm{m})e^{-(2.00 \times 10^{-3}\,\mathrm{s}^{-1})t}\cos(15.4\,\mathrm{s}^{-1}t)$$
$$- (15.4\,\mathrm{s}^{-1})(0.500\,\mathrm{m})e^{-(2.00 \times 10^{-3}\,\mathrm{s}^{-1})t}\sin(15.4\,\mathrm{s}^{-1}t)$$
$$v(t) = -[(1.00 \times 10^{-3}\,\mathrm{m/s})\cos(15.4\,\mathrm{s}^{-1}t) + (7.70\,\mathrm{m/s})\sin(15.4\,\mathrm{s}^{-1}t)]e^{-(2.00 \times 10^{-3}\,\mathrm{s}^{-1})t}$$

Notice that the units in the exponent and in the trig functions will cancel. The units of the velocity are m/s, which is what we got. Checking your units can help you catch algebra mistakes.

EXAMPLE 2.3

Peppy LePew sees his beloved Pussy LeMew walking down the street on a sunny, Saturday morning. He wants to catch her, so he begins to run after her. His position, relative to hers, goes as:

$$s(t) = (7.25\,\mathrm{m/s})t - (1.21\,\mathrm{m/s}^2)t^2 + (25.0\,\mathrm{m}) \qquad (\text{for } t > 0)$$

What is the farthest distance he is away from her? When does he catch her?

SOLUTION

[Given: $s(t) = At^2 + Bt + C$, where $A = -1.21\,\mathrm{m/s}^2$, $B = 7.25\,\mathrm{m/s}$, and $C = 25.0\mathrm{m}$. Find: s_{max} and $t_{s=0}$.] First, to find the farthest distance between Peppy and Pussy, we need to differentiate $s(t)$ to get their relative velocity. When

3

the relative velocity is zero, we have an extreemum (either the farthest or the closest distance). We then use the second derivative to determine if it is a maximum or minimum distance. So,

$$v = \frac{ds(t)}{dt} = \frac{d}{dt}[At^2 + Bt + C]$$
$$v = 2At + B$$

Setting $v = 0$ gives $0 = 2At + B$ or $t = -\frac{B}{2A}$. Plugging in the values of A and B gives:

$$t = -\frac{(7.25\,\text{m/s})}{2(-1.21\,\text{m/s}^2)} = 3.00\,\text{s}$$

To determine if this is a maximum, we need to take the second derivative:

$$\frac{d^2 s(t)}{dt^2} = \frac{dv}{dt} = \frac{d}{dt}(2At + B) = 2A = 2(-1.21\,\text{m/s}^2) = -2.42\,\text{m/s}$$

If $\frac{d^2 s(t)}{dt^2} < 0$, then we have a maximum, so we indeed have a maximum. Now, plugging $t = 3.00\text{s}$ back into $s(t)$ will give us the maximum distance between Peppy and his beloved.

$$s_{\text{max}} = (-1.21\,\text{m/s}^2)(3.00\,\text{s})^2 + (7.25\,\text{m/s})(3.00\,\text{s}) + (25.0\,\text{m})$$
$$= 35.9\,\text{m}$$

So the maximum separation is *35.9m.*

Next, we need to determine when he catches her. He will catch her when $s(t) = 0$, so:

$$0 = At^2 + Bt + C$$

4

To solve this we need the quadratic formula:

$$t = \frac{-B \pm \sqrt{B^2 - 4AC}}{2A}$$

$$= \frac{-(7.25\,\text{m/s}) \pm \sqrt{(7.25\,\text{m/s})^2 - 4(-1.21\,\text{m/s}^2)(25.0\,\text{m})}}{2(-1.21\,\text{m/s}^2)}$$

$$t = 3.00\,\text{s} \pm \frac{\sqrt{173.5\,\text{m}^2/\text{s}^2}}{-2.42\,\text{m/s}^2} = 3.00\,\text{s} \pm 5.44\,\text{s}$$

$$= -2.44\,\text{s} \quad \text{or} \quad 8.44\,\text{s}$$

Since we want only the positive value, it takes Peppy 8.44s to catch Pussy.

EXERCISE 2.1

A dog is on a leash and being walked by his owner. The dog and his owner stop before crossing the street. If the dog's distance from the corner as a function of time is:

$$\ell(t) = (3.00\,\text{cm/s})t + (4.50\,\text{cm/s}^2)t^2 + (1.70\,\text{cm/s}^3)t^3,$$

what it the dog's speed, one second after it begins to cross the street?

EXERCISE 2.2

A mosquito flies in a straight line. Its position can be described as $x(t) = (50\,\text{cm/s})t + (10\,\text{cm/s}^2)t^2$. (a) What is its average speed during the first 2.0 s of travel? (b) What is its instantaneous speed at $t = 1.5\text{s}$?

EXERCISE 2.3

The driver of a car sees a stop sign 50.0 m ahead and begins to slow down. If his distance from the sign as a function of time is

$$\mathbf{r}(t) = (14.7\,\text{m/s})t - (10.7\,\text{m/s}^2)t^2 \quad \text{— toward the stop sign,}$$

(a) what is his speed at $t = 0$? (b) How long does it take for him to stop?

5

EXERCISE   2.4

Blood flowing through an artery in the human body can be modeled as a particle traveling with its position as a function of time given by:

$$x(t) = (10.0\,\text{cm/s})t + (1.50\,\text{cm})\cos[(1.00\,\text{s}^{-1})t].$$

The first term represents the average blood flow and the second term represents the pulsing of the heart. Calculate the maximum speed of the blood through this artery.

EXERCISE   2.5

A squirrel runs along a telephone wire so that its position from the house is described as:

$$x(t) = (3.0\,\text{m/min}^2)t^2 - (0.50\,\text{m/min}^3)t^3 + (7.5\,\text{m/min})t.$$

(a) What is the squirrel's speed as a function of time? (b) At what time is the squirrel stationary?

EXERCISE   2.6

A dog is connected by a chain to a tree so that it doesn't run away. If the chain is 10 m long and the dog's distance from the tree varies with time as:

$$r(t) = (4.0\,\text{m/s})t + (0.50\,\text{m/s}^2)t^2,$$

what is the radial speed of the dog, just before it gets to the end of its chain?

EXERCISE   2.7

A stick is floating on water in a river. If the position of the stick (relative to its starting point) is

$$s(t) = (0.50\,\text{m/s}^2)t + (2.7\,\text{m/s})t,$$

what is the speed of the stick as a function of time?

EXERCISE 2.8

An airplane starts from rest. If its position varies with time as $s(t) = \frac{1}{3}bt^3$, where $b = 1.00\,\text{m/s}^3$, (a) how long does it take to reach a speed of $100\,\text{km/h}$ ($62\,\text{mi/h}$)? (b) What is its speed after it has traveled $100\,\text{m}$?

EXERCISE 2.9

Two cars move parallel to one another along a road. Car A's position as a function of time can be described as $r_A(t) = 3.0\,\text{m} + (2.5\,\text{m/s})t$, while car B's position is described by $r_B(t) = 7.0\,\text{m} + (11.7\,\text{m/s})t - (2.0\,\text{m/s}^2)t^2$. Write an expression for the speed of car A with respect to car B.

EXERCISE 2.10

Consider an $8.00\,\text{oz}$. paper cup. The radius changes as you go up the cup: $r = 2.5\,\text{cm} + 0.13h$. The cup is initially filled with water to a height of $7.00\,\text{cm}$. A person drinks the water through a straw so that a constant amount of water is removed per second. If the rate of water removal is 15 milliliters per second, calculate the speed at which the waterline is dropping when the height of water is $2.0\,\text{cm}$.

Chapter 3   KINEMATICS: ACCELERATION

EXAMPLE   3.1

A boy scout raises his food pack which has mass m, using a rope that is thrown over a tree limb at a height h above his hands. The idea is to protect the pack from hungry animals that may raid his camp, (see figure EG-3.1a). He walks away from the vertical rope with a constant speed v_0 holding the free end of the rope in his hands so that he can tie it to a nearby rock. (a) Find the speed of the food pack on its way up while the boy is walking away, from the vertical portion of the rope in a straight line. (b) What is the magnitude of the acceleration of his food pack? (c) What is the speed and the magnitude of the acceleration of the pack shortly after the boy leaves the location adjacent to the vertical rope?

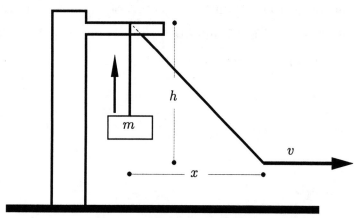

Figure: EG-3.1a

SOLUTION

[Given: m, h, v_0; consult figure EG-3.1a. Find: (a) upward speed of the mass m; (b) upward acceleration of the mass m; (c) speed and the magnitude of the acceleration at the beginning.]

Figure EG-3.1a is essential for solving this problem. Observe that the direction in which the food pack moves and the direction in which the boy scout walks are perpendicular to each other. Denoting the distance of the boy scout from the vertical by x enables us to construct a right triangle depicted in Fig. EG-3.1b. Let us designate the hypotenuse of that triangle as l. It represents the length of the rope from the tree limb to the scout's hand.

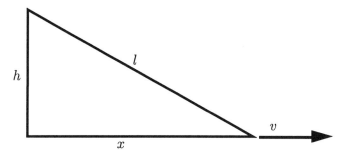

Figure: EG-3.1b

The velocity of the boy scout is also shown in Fig. EG-3.1b. The key relationship in the solution of this problem is the expression for the length l of the rope from the tree limb to the boy scout's hand; It is given by Pythagorean Theorem which holds for any right triangle. Accordingly, $l = \sqrt{x^2 + h^2}$. While the boy walks away from the vertical the distance x increases at constant rate (it is a linear function of time). However, the food pack will accelerate upward toward the tree limb because its height depends on l (as l gets longer the pack's height increases even though h is constant) which is not a linear function of time.

(a) Let us assume that the boy scout is a distance x away from the vertical. Since he is moving with constant speed v_0, during a small time interval Δt he travels a horizontal distance $\Delta x = v_0 \Delta t$. This increases the length l of the rope by a distance Δl which is precisely the amount by which the food pack rises during the time interval Δt. In other words, the upward speed of the pack is given as the time derivative of the l, $v_p = dl/dt$. We have

$$v_p = \frac{d}{dt}\sqrt{x^2 + h^2} = \frac{x}{\sqrt{x^2 + h^2}}\frac{dx}{dt} \; .$$

Since $dx/dt = v_0$, the speed at which the pack rises is given by

$$v_p(t) = \frac{xv_0}{\sqrt{x^2 + h^2}} = \frac{v_0^2 t}{\sqrt{v_0^2 t^2 + h^2}} \; ,$$

where we used $x = v_0 t$ in the second equality in order to express the speed v_p of the food pack in terms of time t.

(b) From the answer to part (a) we see that the upward speed of the food pack depends on the horizontal distance x which is itself changing linearly with time. However, since the increase in the height of the pack depends on the length l of the rope, which is not a linear function of time, the speed of the food pack is a nonlinear function of time. This means that the pack is experiencing an acceleration. To find that acceleration we calculate the time derivative of the speed v_p. Note that the denominator of the expression for the speed v_p, namely $\sqrt{x^2 + h^2}$, precisely equals l! Thus

$$a_p = \frac{dv_p}{dt} = \frac{d}{dt}\frac{v_0^2 t}{\sqrt{v_0^2 t^2 + h^2}} = \frac{d}{dt}\left(\frac{v_0 t}{l}\right)$$

$$= \frac{v_0^2}{l} + v_0^2 t\frac{d}{dt}\frac{1}{l} = \frac{v_0^2}{l} - v_0^2 t\frac{1}{l^2}\frac{dl}{dt} = \frac{v_0^2}{l} - v_0^2 t\frac{v_p}{l^2} \; .$$

Using, $x = v_0 t$ and $v_p = xv_0/\sqrt{x^2 + h^2} = xv_0/l$ we can write the acceleration a_p in terms of v_0, h, and l, the given quantities:

$$a_p = \frac{v_0^2}{l} - v_0 x\frac{v_0 x}{l^3} = \frac{v_0^2}{l}\left(1 - \frac{x^2}{l^2}\right) = \frac{v_0^2 h^2}{l^3} \; ,$$

and in writing the last equality we have used the relation $l^2 = x^2 + h^2$. Hence,

$$a_p = \frac{v_0^2 h^2}{(x^2 + h^2)^{3/2}} = \frac{v_0^2 h^2}{\left(v_0^2 t^2 + h^2\right)^{3/2}} \; .$$

(c) To find the speed and the scalar value of the acceleration at the beginning, we

10

evaluate the results obtained in parts (a) and (b) at time $t = 0$. At time $t = 0$, the boy is at the vertical rope, $x(0) = 0$. Hence $v_p(0) = 0$, and $a_p(0) = v_0^2/h$. Notice that the acceleration has the correct dimension – unit of distance divided by the square of the unit of time!

EXAMPLE 3.2

The acceleration of a marble moving in some oily liquid is proportional to the square of its speed: $a = -bv^2$, where $b = 2.0\,\mathrm{m^{-1}}$ and $v > 0$. If the marble enters the liquid with a speed of $2.0\,\mathrm{m/s}$ how long will it take before its speed is reduced to half of its initial value?

SOLUTION

[Given: $a = -bv^2$, where $b = 2.0\,\mathrm{m^{-1}}$, and $v > 0$; $v_0 = 2.0\,\mathrm{m/s}$. Find: t when $v = v_0/2 = 1.0\,\mathrm{m/s}$.]

The acceleration is the time derivative of the velocity. Hence, we must integrate the acceleration over the time interval during which the motion takes place to find the speed. Accordingly,

$$\int_{v_0}^{v} dv = \int_{0}^{t} a(t)\ dt\ .$$

The above integral relation assumes that we know acceleration as a function of time and may or may not be the case. In the problem at hand we know that the acceleration is a quadratic function of the speed, but we don't know the explicit time dependence of the acceleration. Yet we can find the solution to our problem. Note that the acceleration equation has the form

$$a = \frac{dv}{dt} = -bv^2\ .$$

There are two variables here (v and t) and we must group them so that all the v-terms are on one side and all the t-terms are on the other. Dividing the equation

11

by the square of the speed and by multiplying it by dt we obtain a form that can be integrated. Therefore,

$$\int \frac{1}{v^2} \frac{dv}{dt} dt = \int \frac{dv}{v^2} \qquad \text{whereas} \qquad \int \frac{-bv^2}{v^2} dt = -b \int_0^t dt$$

and so

$$\int_{v_0}^{v} \frac{dv}{v^2} = -b \int_0^t dt \; .$$

note the limits; when $t = 0$, $v = v_0$, when $t = t$, $v = v$. These indefinite integrals are now easy to evaluate,

$$\int_{v_0}^{v} \frac{dv}{v^2} = \left[-\frac{1}{v} \right]_{v_0}^{v} = \frac{1}{v_0} - \frac{1}{v} = -bt$$

and $v(t) = v_0/(1 + btv_0)$. Since we need to find the time when the speed of the marble is reduced to half of its initial value, set $v = v_0/2$ at time t:

$$\left[-\frac{1}{v} \right]_{v_0}^{v_0/2} = -bt$$

$$= \frac{1}{v_0} - \frac{1}{v_0/2} = -\frac{1}{v_0} = -bt$$

and this is readily solved for time:

$$t = \frac{1}{bv_0} = \frac{1}{(2.0\,\text{m}^{\text{-1}})(2.0\,\text{m/s})} = 0.25\,\text{s} \; .$$

The time it takes the marble to loose half of its initial speed $(2.0\,\text{m/s})$ in the oily liquid is $0.25\,\text{s}$. Note that the final answer is given to two digit precision – same as the precision of all other quantities in this problem.

EXAMPLE 3.3

A ball is thrown vertically upward with initial speed v_0. In addition to the gravitational acceleration the ball also experiences an acceleration due to air resistance equal to $a_r = -bv$. The negative sign of the resistive acceleration indicates that the acceleration is in the direction opposite of the direction of motion. Find the maximum height that the ball will reach. How long will it take the ball to reach the maximum height? Also show that in the limit when the air resistance is not present the results reduce to the known results for the free fall!

SOLUTION

[Given: initial upward speed v_0, acceleration due to gravity, $g = -9.81 \, \mathrm{m/s^2}$ pointing downward, resistive acceleration $a_r = -bv$, pointing in the direction opposite of the direction of motion. Find: maximum height that the ball can achieve; show that when the air resistance is absent the results reduce to free-fall case.]

For definiteness let us take the positive direction to be upward. Then the gravitational acceleration which is negative $a_g = g$, points down. The ball will travel upward which means that the resistive acceleration is also pointing downwards and that we can write $a_r = -bv$. Note that if the motion was in the downward direction and we keep the upward direction positive, the sign of the gravitational acceleration would remain the same but we would have to change the sign of the resistive acceleration. The good starting point to find the solution to this problem is the acceleration equation,

$$ a = \frac{dv}{dt} = a_g + a_r = g - bv \ . $$

This equation immediately tells us that both, gravity and air slow the ball down. The slowdown also happens when only the gravity acts and the ball is thrown upward, but with the resistance included, the slowdown is quicker. Notice that if we set $b = 0$ we have the case of no resistance. This will provide a nice way to check

13

that the final result makes sense.

The acceleration equation can not be integrated over the time directly because we don't know the explicit time dependence of the acceleration. However, we can write $dv/bv - g = -dt$ which can readily be integrated. Using the integral formula $\int dz/(a + bz) = \frac{1}{b}\ln(a + bz)$ it follows that

$$\int\limits_{v_0}^{v} \frac{dv}{bv - g} = -\int\limits_{0}^{t} dt$$

and so

$$\frac{1}{b} \ln \frac{vb - g}{v_0 b - g} = -t \ .$$

To solve the resulting formula for v exponentiate the result we obtained:

$$\frac{bv - g}{bv_0 - g} = \mathrm{e}^{-bt} \ .$$

Multiplying the equation by the denominator of the left hand side, $(bv_0 - g)$, yields

$$bv - g = \mathrm{e}^{-bt}(bv_0 - g) \quad \text{and} \quad v = \frac{g}{b}\left(1 - \mathrm{e}^{-bt}\right) + v_0\mathrm{e}^{-bt} \ .$$

In writing the expression for $v(t)$ the terms are grouped in a way that is easier to understand and is more convenient for some of the calculations that follow. At this point it is a good idea to verify that the result makes sense in the case when no air resistance is present. To do that consider the case when $b = 0$. The term $v_0\mathrm{e}^{-bt}$ simply becomes v_0 – exactly what is expected in the velocity equation when only gravity acts. In the term $g(1 - \mathrm{e}^{-bt})/b$ the limit is taken with great care because the constant b appears in the denominator. Note that for small b we can write[*]

---

[*] Consult your calculus book on the subject of expansion of exponential function into a power series.

$1 - e^{-bt} = 1 - (-bt + \frac{1}{2}b^2t^2 + \ldots) \approx bt - \frac{1}{2}b^2t^2$, where the terms with higher powers of b have been neglected. Dividing this result by b yields $g(1 - e^{-bt})/b \approx gt - \frac{1}{2}gbt^2$ and b can be safely set to zero. As a matter of fact, all the terms initially neglected will become zero in the limit! This leaves us with the result $v \rightarrow gt + v_0$ when $b \rightarrow 0$. This is just what we expect to have when air resistance is not present.

There is another limit which is quite interesting to analyze. Consider the case when the ball spends a very long time traveling through the air. We are considering the limit when $t \rightarrow \infty$. In this limit $e^{-bt} \rightarrow 0$. Then the scalar value of the velocity becomes $v \rightarrow \frac{g}{b}$. Note that the scalar value of the velocity is negative which tells us that this limit occurs after the ball has reached its peak height and is on its way back toward the ground. Note also, that the velocity in this limit is constant. This phenomenon is known as terminal speed. On the way down, the resistance and gravity act in opposite directions. At some point, the rate at which the speed is increased by the gravity exactly matches the rate at which the resistance reduces the speed.

Having found the speed as a function of time we can now compute the position. Since $v = dy(t)/dt$ and the speed is given as a function of time, we can carry out the integration to get $y(t)$. With the ground level as the coordinate origin,

$$\int_0^y dy = y(t) = \int_0^t v(t)\,dt = \int_0^t \left[\frac{g}{b} + \left(v_0 - \frac{g}{b} \right) e^{-bt} \right] dt \ .$$

The integration is not difficult (use the integral formula $\int dz\ e^{cz} = \frac{1}{c}e^{cz}$) and it yields

$$y(t) = \frac{gt}{b} \int_0^t dt + \left(v_0 - \frac{g}{b} \right) \int_0^t dt\ e^{-bt}$$

$$y(t) = \frac{gt}{b} + \left(v_0 - \frac{g}{b} \right) \frac{1}{b}(e^{-bt} - 1) \ .$$

This looks like a complicated expression and to some extent it is! The first order of business is to verify that it makes sense. In the limit as $b \to 0$, $y(t)$ should reduce to the known result for the free-fall motion. As in the case of the speed, the limit has to be taken carefully. As before, we use the expansion of the exponential function into a series. However, this time it is necessary to consider the terms that are quadratic in b because in the expression for $y(t)$ there are divisions by b^2. hence, write $e^{-bt} - 1 \approx -bt + \frac{1}{2}b^2t^2$. The terms neglected will have higher powers of b and in the limit they vanish. Then, in the limit as $b \to 0$

$$
\begin{aligned}
y &\to \frac{gt}{b} + \left(v_0 - \frac{g}{b}\right)\frac{1}{b}\left(bt - \tfrac{1}{2}b^2t^2\right) \\
&= \frac{gt}{b} + \left(v_0 - \frac{g}{b}\right)\left(t - \tfrac{1}{2}bt^2\right) \\
&= \frac{gt}{b} - \frac{gt}{b} + v_0 t + \frac{gt^2}{2} - \frac{bv_0t^2}{2} \\
&= v_0 t + \tfrac{1}{2}gt^2 \ .
\end{aligned}
$$

But this is exactly what is expected.

Next we begin with the final part of the work – the calculation of the maximum altitude. At this point this may seem complicated but don't despair. If you have read the problem carefully you remember that we are after the maximum height that can be achieved. This means that we need to evaluate the $y(t)$ at the time instant t when the speed is zero (as in the case of the simple vertical shot). This condition is going to bring some simplifications into our result.

Let t be the time at which the ball has reached its maximum height. At that moment the speed of the ball is zero and we have the condition:

$$
0 = \frac{g}{b}(1 - e^{-bt}) + v_0 e^{-bt} \ .
$$

Thus

$$
e^{-bt} = \frac{-g/b}{v_0 - g/b} \ .
$$

By taking the logarithm we can calculate the time it takes to reach the maximum height,

$$t = \frac{1}{b}\ln(1 - v_0 b/g) = \frac{1}{b}\ln(1 - v_0/v_t) \, ,$$

where $v_t = -g/b$. The value of the maximum height is given by the formula

$$
\begin{aligned}
y &= \frac{g}{b}\frac{1}{b}\ln(1 - v_0 b/g) + \frac{1}{b}\left(v_0 - \frac{g}{b}\right)\left(1 + \frac{g/b}{v_0 - g/b}\right) \\
&= \frac{v_0}{b} + \frac{g}{b^2}\ln(1 - v_0 b/g) \, .
\end{aligned}
$$

Finally, we should check the limit when b vanishes. From your calculus text you need to find the expression for the series expansion of the logarithmic function: $\ln(1 + x) \approx x - \frac{1}{2}x^2 + \ldots$ for $0 < x < 1$. Using this (keeping the terms quadratic in parameter b),

$$
\begin{aligned}
y &\rightarrow \frac{v_0}{b} + \frac{g}{2}\left(-\frac{v_0 b}{g} - \frac{v_0^2 b^2}{2g^2}\right) \\
&= \frac{v_0}{b} - \frac{v_0}{b} - \frac{v_0^2}{2g} = -\frac{v_0^2}{2g} \, .
\end{aligned}
$$

Indeed the limit is what we expect.

EXAMPLE 3.4

A ball is thrown straight up into the air and its height is described by the expression $h(t) = At + Bt^2$, wherein $A = 19.6\,\mathrm{m/s}$ and $B = -4.90\,\mathrm{m/s^2}$. (a) Find the average speed and the average velocity of the ball over a time interval $t = 0.00\,\mathrm{s}$ to $t = 4.00\,\mathrm{s}$. (b) Find the average speed and the average velocity of the ball over a time interval $t = 3.00\,\mathrm{s}$ to $t = 4.00\,\mathrm{s}$. (c) Calculate the instantaneous speed as a function of time and find the speed and velocity at times $t = 0.00\,\mathrm{s}, 1.00\,\mathrm{s}, 2.00\,\mathrm{s}, 3.00\,\mathrm{s}$ and $4.00\,\mathrm{s}$.

SOLUTION

[Given: $h(t) = At + Bt^2$, $A = 19.6\,\mathrm{m/s}$, $B = -4.90\,\mathrm{m/s^2}$ Find: $v_{av}(t)$, $\mathbf{v}_{av}(t)$ and the speed and velocity at times $t = 0.00\,\mathrm{s}, 1.00\,\mathrm{s}, 2.00\,\mathrm{s}, 3.00\,\mathrm{s}$ and $4.00\,\mathrm{s}$.]

Before we begin the problem, lets plot the height as a function of time. To prepare for plotting the function in question we create a table of function values in the interval of interest.

t (s)	0.00	0.500	1.00	1.50	2.00	2.50	3.00	3.50	4.00
$h(t)$ (m)	0.00	8.60	14.7	18.4	19.6	18.4	14.7	8.60	0.00

The graph of distance as a function of time is given in Fig. EG-3.4.

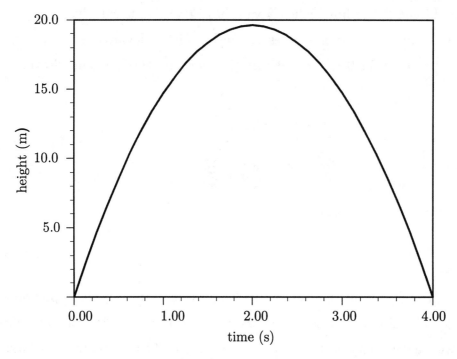

Figure: EG-3.4

(a) First we want to find the average speed over the whole trip. Using Eq (2.3) on page 34 of the text, $v_{av} = \Delta l / \Delta t$, where Δl is the total distance traveled and Δt is the time elapsed. Therefore, Δl is the distance to the top plus the distance back down, or, from the curve: $\Delta l = 19.6\,\mathrm{m} + 19.6\,\mathrm{m} = 39.2\,\mathrm{m}$ and $\Delta t = 4.00\,\mathrm{s} - 0.00\,\mathrm{s} = 4.00\,\mathrm{s}$. So

18

$$v_{av} = \frac{39.2\,\text{m}}{4.00\,\text{s}} = 9.80\,\text{m/s} \ .$$

Next we want to find the average velocity over the whole trip, $\mathbf{v}_{av}$. The answer is different from the average speed. Since $\mathbf{v}_{av} = \mathbf{s}/t$, where $\mathbf{s}$ is the resultant displacement of the ball, $\mathbf{s} \neq \Delta l$. The ball starts at $h = 0$ and ends at $h = 0$, so $\mathbf{s} = 0$. In terms of vectors, $(19.6\,\text{m}(-\,\text{up}))$ cancels $(19.6\,\text{m}(-\,\text{down}))$ and $\mathbf{s} = 0.00\,\text{m}$. So we see that there is a big difference between speed and velocity. In calculating average speed we do not care what direction you traveled, we care only *that* you traveled. On the other hand, the average velocity involves the net displacement you experienced. Since the ball ended up where it began, the average *velocity* is zero; since the ball moved, the average *speed* is nonzero.

(b) Now that we understand the difference between speed and velocity, let's look at the last second of travel. The average speed is $v_{av} = \Delta l / \Delta t$ and $\Delta l = |l_f - l_i| = |0.00\,\text{m} - 14.7\,\text{m}| = 14.7\,\text{m}$ (we only want the distance traveled). Since $\Delta t = 4.00\,\text{s} - 3.00\,\text{s} = 1.00\,\text{s}$, we have that the average speed is $v_{av} = 14.7\,\text{m}/1.00\,\text{s} = 14.7\,\text{m/s}$. So, the ball travels faster at the end of the trip than it did on the average for the whole trip.

What about the average velocity? Since $\mathbf{v}_{av} = \Delta \mathbf{s}/\Delta t$, we need to find the displacement during the last second of the trip. Measuring up from the origin, we have $\mathbf{s} = \mathbf{s}_f - \mathbf{s}_i = 0 - (14.7\,\text{m}(-\,\text{up}))$, where $(-14.7\,\text{m}(-\,\text{up})) = (14.7\,\text{m}-\,\text{down})$. So, the average velocity is

$$\mathbf{v}_{av} = \frac{(14.7\,\text{m}-\,\text{down})}{1.00\,\text{s}} = 14.7\,\text{m/s}-\,\text{down} \ .$$

In this particular instance, the difference between the average speed and the average velocity is only that the there is a direction associated with average velocity (*down* in this case).

(c) Now that we understand how to calculate average speeds and velocities, what if we keep reducing the time interval so that it approaches zero? Then we can investigate speeds and velocities at an exact point in time; we get instantaneous speeds and velocities. The velocity, we now understand, has a direction associated with it: it is a vector. The speed does not: speed is a scalar. Since instantaneous velocity is $\mathbf{v} = d\mathbf{s}/dt$ and instantaneous speed is $v = dl/dt$, the instantaneous speed is just the magnitude of the instantaneous velocity. To get the instantaneous speed, we need to differentiate $h(t)$, with respect to time.

$$v = \frac{dh}{dt} = \frac{d}{dt}\left(At + Bt^2\right) = A + 2Bt \ .$$

Since constants A and B are known we have

$$v = 19.6\,\text{m} - 9.80t \ .$$

In addition the instantaneous speed is just the magnitude of the velocity, so we can calculate the velocity too. Let's tabulate the results:

time	velocity	speed
t (s)	$\mathbf{v}$ (m/s)	v (m/s)
0	19.6 – up	19.6
1	9.80 – up	9.8
2	0.00	0
3	9.80 – down	9.8
4	19.6 – down	19.6

But wait! Have we done something wrong? We got $v = 0\,\text{m/s}$ at $t = 2.00\,\text{s}$; that can't be right, can it? Yes it can. At $t = 2.00\,\text{s}$, the ball just reaches the top of its flight and at that instant it has zero velocity. Notice on the graph we made at the beginning of the problem that the slope is zero at $t = 2.00\,$s. The speed corresponds to the slope, and at $t = 2.00\,$s, $v = 0$: the tangent to the curve in horizontal.

EXERCISES

EXERCISE 3.1

An ant moving in a plane has velocity components given by

$$v_x = \alpha + \beta t \qquad \text{and} \qquad v_y = \gamma + \delta t ,$$

where $\alpha, \beta, \gamma,$ and δ are constants. (a) Using dimensional analysis give the interpretation of constants in the defining relations for the velocity. (b) Find the position of the ant as a function of time. (c) Find the acceleration of the ant. (d) Find the magnitude and the direction of the velocity and acceleration vectors.

EXERCISE 3.2

A car travels across a flat bottom of a salt lake and its instantaneous position relative to some coordinate system is given by:

$$x = x_0 + b_x t + c_x t^2 + d_x t^3 ,$$
$$y = y_0 + b_y t + c_y t^2 + d_y t^3 ,$$

where $x_0, b_x, c_x, d_x, x_0, b_y, c_y$ and d_y are constants. Assume that the car starts at time $t = 0$. Find expressions for the initial velocity and acceleration components in the x and y directions.

EXERCISE 3.3

A marble moves along a straight line and experiences a constant acceleration a. Assume that at the initial moment the scalar value of the velocity of the marble was v_0 and that at the final moment the scalar value of the velocity was v. (a)

Show, using calculus that the average scalar value of the velocity of the marble is given by the formula $v_{av} = \frac{1}{2}(v_0 + v)$. (b) Does the formula for the average scalar value of the velocity given in part (a) works in the case when the acceleration is not constant? To help you get a correct answer consider the case of the marble described in this problem and assume that at time t the acceleration ceases to act and that you continue to observe the marble for an additional time interval t_a. Calculate the average scalar value of the velocity of the marble in this case. (HINT: A time average of some physical quantity $G(t)$ over a time interval from t to $t + \Delta t$ is defined as $G_{av} = \frac{1}{\Delta t} \int_t^{t+\Delta t} G(t) \, dt$.)

EXERCISE 3.4

A small object starts at $t = 0$ from the origin with speed $v_0 = 0.10 \, \text{m/s}$ and moves along a straight line. It experiences an acceleration which is proportional to the distance from the origin, in particular, $a = -\alpha x$, where $\alpha = 40 \, \text{s}^{-2}$. How far from the origin is the object going to be when its speed becomes zero?

EXERCISE 3.5

A bumble bee while in flight at fixed altitude, accelerates horizontally along a straight line according to $a = \alpha t$, where α is a constant. Find the displacement of the bumble bee as a function of time.

EXERCISE 3.6

A ball is moving along a straight line path and its scalar value of velocity, as a function of time, is given by

$$v(t) = -(1 \, \text{ms}^{-3})t^2 + (8 \, \text{ms}^{-2})t.$$

Consider the time interval from $t = 0 \, \text{s}$ to $t = 10 \, \text{s}$. Find (a) the range of times when the speed is increasing and the range when it is decreasing, (b) the instant of time when the velocity is maximum and its value at the maximum, and (c) the instant of time when the speed is maximum and its value at the maximum.

EXERCISE 3.7

Find the velocity and acceleration of an object moving in space if the components of the position vector are given as a functions of time by

$$x(t) = (8\,\text{m/s})t\ ,$$
$$y(t) = (20\,\text{m/s}^2)t^2,$$
$$z(t) = (24\,\text{m})\ .$$

What is the trajectory of the object?

EXERCISE 3.8

Find the velocity and acceleration of an object moving in a plane if the components of the position vector are given as a functions of time by

$$x(t) = (12\,\text{m})\sin(15\,\text{s}^{-1})t\ ,$$
$$y(t) = (25\,\text{m/s})t\ .$$

EXERCISE 3.9

A small particle suspended in a viscous liquid has its position given by $x = 10e^{-0.2t}$ m, where time t is measured in seconds. (a) Plot the position as a function of time for t between 0 and 10 s. (b) Find the average velocity between 2 and 3 seconds. (c) Find the acceleration of this particle.

EXERCISE 3.10

Find the velocity and the acceleration of an object moving in a plane that has its position given as a function of time by

$$x(t) = e^{(2\,\text{s}^{-1})t}, \qquad z(t) = (4.0\,\text{cm/s}^2)t\ .$$

23

Chapter 4   NEWTON'S THREE LAWS: MOMENTUM

EXAMPLE   4.1

A soda can is set on a large block of ice (so there is no friction between the can and the ice). Someone shoots a 3.0 g BB at the can, which penetrates the can. The velocity of the BB after it enters the can is described by

$$v = 2.0\,\text{cm/s} - (1.0\,\text{cm/s}^3)t^2,$$

until it stops. (a) Calculate the force on the BB due to the can and soda. (b) Calculate the impulse delivered to the BB from the can and soda during the first 0.50 seconds.

SOLUTION

[Given: $v = A - Bt^2$, $A = 2.0\,\text{cm/s}$, $B = 1.0\,\text{cm/s}^3$. Find: (a) $F(t)$, and (b) impulse for time interval $0\,\text{s} < t < 0.50\,\text{s}$.] (a) To calculate the force, remember that $F = \frac{dp}{dt}$ and $p = mv$. So, we need to calculate the change in momentum with time. The mass is constant, but the velocity is changing, so $\frac{dp}{dt} = m\frac{dv}{dt}$. So,

$$F = \frac{dp}{dt} = m\frac{dv}{dt} = m\frac{d}{dt}\left(A - Bt^2\right) = -2mBt$$
$$= -2(3.0\,\text{g})(1.0\,\text{cm/s}^3)t$$
$$= -(6.0\,\text{g cm/s}^3)t\frac{1\,\text{kg}}{1000\,\text{g}}\frac{1\,\text{m}}{100\,\text{cm}} = -(6.0\times10^{-5}\,\text{N/s})t.$$

Note, a negative force means it is slowing the BB down. (b) To calculate the impulse, remember impulse $= \mathbf{p}_f - \mathbf{p}_i = \int_{t_i}^{t_f} F(t)\,dt$. So, since we have calculated

24

the force, we can integrate it.

$$\text{impulse} = \int_0^{0.5\,\text{s}} F(t)\ dt = \int_0^{0.5\,\text{s}} (-6.0 \times 10^{-5}\,\text{N/s})\ dt = \frac{1}{2}(-6.0 \times 10^{-5}\,\text{N/s})t^2\Big|_0^{0.5\,\text{s}}$$
$$= -(3.0 \times 10^{-5}\,\text{N/s})(0.5\,\text{s})^2 - 0$$
$$= -7.5 \times 10^{-6}\,\text{N s}\ .$$

The impulse is negative, which tells us the momentum is getting smaller (and in this case the velocity is getting smaller). Now that we have done the problem using the force, you should notice that we could have done part (b) another way. Since the impulse is the change in the momentum, we can use the momentum directly $p_i = mv_i$, $p_f = mv_f$ where $t_i = 0$ and $t_f = 0.5\,\text{s}$, so

$$\text{impulse} = mv_f - mv_i = m(v_f - v_i) = m\left\{\left[A - B(0.5\,\text{s})^2\right] - \left[A - B(0\,\text{s})^2\right]\right\}$$
$$= m(-B)(0.5\,\text{s})^2$$
$$= -(3.0\,\text{g})(1.0\,\text{cm/s}^3)(0.5\,\text{s})^2\frac{1\,\text{kg}}{1000\,\text{g}}\frac{1\,\text{m}}{100\,\text{cm}}$$
$$= -7.5 \times 10^{-6}\,\text{N s.}$$

I would recommend this second method on an exam because if you do part (a) wrong, you won't use that in part (b).

What happens after the BB punctures the can? Well, the BB will create the hole, and soda will begin to pour out. If no soda poured out of the hole, then the can (with BB inside) would travel at a constant velocity because there is no friction acting in the problem. What happens to the velocity of the can if soda pours out of the hole with constant velocity? Does the can then travel faster or slower? This question requires us to use another concept discussed in this chapter: conservation of momentum. After the BB is inside the can, the can and the BB have the momentum that the BB started with: $\mathbf{p}_i = m_{BB}\mathbf{v}_{BB}$, $\mathbf{p}_f = (m_{BB} + m_{SODA})\mathbf{v}_{BB\&SODA}$. So the can and BB move at constant velocity, which is slower than the BB initially.

Once the soda starts squirting out, it takes some momentum away from the can and BB. Actually, since the momentum taken away is in the opposite direction to the motion of the can, the magnitude of the can's momentum in the original direction of the bullet increases. Since the mass is decreasing, the velocity must increase to compensate for the decreased mass and increased momentum. So, the can travels *faster* as soda squirts out!

EXAMPLE 4.2

The driver of a car sees a stop sign 100 ft ahead and begins to slow down. The velocity of the car, as a function of time, is

$$\mathbf{v}(t) = (40\,\mathrm{mi/h})e^{-(0.25\,\mathrm{s}^{-1})t} - 10\,\mathrm{mi/h}.$$

Note that this velocity $v(t)$ is valid only for times when $v(t)$ is nonnegative. Calculate the force as a function of time (due to the brakes, friction, *etc.*) which is needed to stop the car. Assume the mass of the car is $1,000\,\mathrm{kg}$.

SOLUTION

[Given: $v(t) = Ae^{-Bt} - C$, $A = 40\,\mathrm{mi/h} = 17.9\,\mathrm{m/s}$, $B = 0.25\,\mathrm{s}^{-1}$, $C = 10\,\mathrm{mi/h} = 4.47\,\mathrm{m/s}$, $m = 1,000\,\mathrm{kg}$. Find: $F(t)$.] In this problem the mass of the car is constant, but the velocity is changing. So since $F = \frac{dp}{dt} = \frac{d(mv)}{dt}$, $F = m\frac{dv}{dt}$. So, we need to differentiate the velocity to calculate the force:

$$F = m\frac{dv}{dt} = m\frac{d}{dt}\left(Ae^{-Bt} - C\right) = -\,AmBe^{-Bt}$$
$$= (1000\,\mathrm{kg})(-17.9\,\mathrm{m/s})(0.25\,\mathrm{s}^{-1})e^{-(0.25\,\mathrm{s}^{-1})t}$$
$$= -\,(4475\,\mathrm{kg\,m/s^2})e^{-(0.25\,\mathrm{s}^{-1})t}.$$

So the force needed is $F = -(4.5 \times 10^3\,\mathrm{N})\exp\left(-(0.25\,\mathrm{s}^{-1})t\right)$. The negative sign says the force is slowing the car down. Just for a check, let's calculate how long it takes to stop the car. This actually doesn't require us to use any calculus, and we don't need to use the expression for the force, but it allows us to see if we are stopping in approximately the right amount of time. The car stops when $v(t) = 0$, so:

$$v(t) = 0\,\text{m/s} = (17.9\,\text{m/s})e^{-(0.25\,\text{s}^{-1})t} - 4.47\,\text{m/s}$$
$$\Rightarrow 0.25 = e^{-(0.25\,\text{s}^{-1})t}$$
$$\ln 0.25 = -(0.25\,\text{s}^{-1})t$$
$$t = 5.5\,\text{s}.$$

It takes 5.5 s for the car to come to a stop. (That seems reasonable.) Notice, however, that as the velocity approaches zero, the force does not approach zero; in fact, it approaches -1100N. The force will be zero *after* the car has stopped, but just as it is stopping, the force is still required!

EXAMPLE 4.3

Consider an open-top railroad car that has high, watertight side walls. The railroad car has mass of 2.00×10^3 kg and is traveling at a speed of 3.00 m/s. For this entire problem, assume there is no friction between the railroad car and the tracks. (a) Calculate the force needed to keep the car moving at 3.00 m/s (b) At time $t = 0.00$ h, it begins to rain. Water starts collecting in the railcar and the mass of the car increases at a constant rate. The mass as a function of time can be expressed as $m(t) = (2.00 \times 10^3\,\text{kg}) + (1.00\,\text{kg/h})t$. During the period of steady rain, calculate the force needed to keep the car moving at 3.00 m/s. (c) After 10 hours the rain starts coming down faster and faster. The mass of the car no longer increases linearly with time. It now has the form:

$$m(t) = (2.00 \times 10^3\,\text{kg}) + (1.00\,\text{kg/h})t + (0.50\,\text{kg/h}^2)(t - 10\,\text{h})^2.$$

During this period of increasing rain, calculate the force (as a function of time) needed to keep the car moving at 3.00 m/s.

SOLUTION

[Given: (a) $m = 2.00 \times 10^3$ kg, (b) $m(t) = m_0 + At$, (c) $m(t) = m_0 + At + B(t - 10\,\text{h})^2$, where $m_0 = 2.00 \times 10^3$ kg, $A = 1.00$ kg/h, and $B = 0.50$ kg/h²; $v = 3.00$ m/s. Find: $F(t)$.] To calculate the force, we need to use Newton's second law, $F = \frac{dp}{dt}$, where $p = mv$ is the momentum. Often, the mass of the system is constant and the velocity changes with time, reducing Newton's second law to $F = ma$. However, in this problem, the velocity remains constant while the mass changes with time (in parts b and c). (a) For this part of the problem, $p = mv$ is constant (both m and v are constant). So, $F = \frac{dp}{dt} = 0$. In other words, no applied force is necessary to keep the train moving. The reason we usually need the engine is because with friction between the train and the track, the answer of $F = 0$ is no longer true.

(b) In this part, the velocity is kept constant, but the mass is changing. So, $\frac{dp}{dt} = \frac{d(mv)}{dt} = \left(\frac{dm}{dt}\right)v + m\left(\frac{dv}{dt}\right)$ or $\frac{dp}{dt} = \left(\frac{dm}{dt}\right)v$. So, $\frac{dm}{dt} = \frac{d}{dt}(m_0 + At) = A$ and so $F = \frac{dp}{dt} = Av = (1.00\,\text{kg/h})(3.00\,\text{m/s})$. Notice that the units are not correct to have the force in newtons. We need to convert the hours into seconds:

$$F = 3.00 \frac{\text{kg} \cdot \text{m}}{\text{hr} \cdot \text{s}} \cdot \frac{1\,\text{hr}}{3600\,\text{s}} = 8.33 \times 10^{-4}\,\text{kg}\,\text{m/s}^2 = 8.33 \times 10^{-4}\,\text{N},$$

Now a force *is* required to keep the train moving at a constant velocity. If the force were not applied, the train would eventually stop.

(c) In this part, as in part b, the velocity is constant and the mass is changing.

$$m = m_0 + At + B(t - 10\,\text{hr})^2$$
$$= m_0 + At + B[t^2 - (20\,\text{hr})t + 100\,\text{hr}^2]$$

28

Now, we can differentiate m to get F.

$$F = \frac{dp}{dt} = \frac{dm}{dt} \cdot v = \frac{d}{dt}[m_0 + At + B(t^2 - 20\,\text{hr} \cdot t + 100\,\text{hr}^2)] \cdot v$$
$$= [A + B(2t - 20\,\text{hr})] \cdot v = [1.0\,\text{kg/hr} + 0.5\,\text{kg/hr}^2(2t - 20\,\text{hr})] \cdot 3.00\,\text{m/s}$$
$$= 3.00\frac{\text{kg} \cdot \text{m}}{\text{hr} \cdot \text{s}} + \left(3.00\frac{\text{kg} \cdot \text{m}}{\text{hr}^2 \cdot \text{s}}\right)(t - 10\,\text{hr})$$
$$= \left(3.00\frac{\text{kg} \cdot \text{m}}{\text{hr}^2 \cdot \text{s}}\right)(t - 9\,\text{hr}) \qquad (\text{for } t \geq 10\,\text{hr})$$

Now, we can convert one of the hrs in the denominator into seconds:

$$F = 3.00\frac{\text{kg} \cdot \text{m}}{\text{hr} \cdot \text{hr} \cdot \text{s}}\left(\frac{1\,\text{hr}}{3600\,\text{s}}\right)(t - 9\,\text{hr})$$
$$= (8.33 \times 10^{-4}\,\text{N/hr})(t - 9\,\text{hr})$$

So, if t is in hours, we get a force in newtons. Notice *now* we need a time-dependent (increasing) force to keep the train moving at constant velocity. So, to summarize:

$$F = \begin{cases} 0 & \{t < 0\} \\ 8.33 \times 10^{-4}\,\text{N} & \{0 < t < 10\,\text{h}\} \\ 8.33 \times 10^{-4}\,\text{N/hr}(t - 9\,\text{hr}) & t > 10\,\text{h} \end{cases}$$

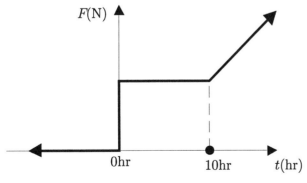

Figure: EG-4.3

EXERCISE 4.1

A block moves, without friction, under the influence of a force described by:

$$F(t) = (5.00\,\text{N/s}^2)t^2 + (3.00\,\text{N/s})t + 4.75N$$

for exactly three seconds. Find the impulse delivered to the block.

EXERCISE 4.2

Imagine you are walking your Great Dane Beowoof on a leash. He spots a squirrel and chases after it, thus pulling you along. For the next one second, your velocity is described by:

$$v(t) = 1.00\,\text{m/s} + (2.00\,\text{m/s}^2)t^2 + (2.00\,\text{m/s}^3)(1.5\,\text{s} - t)^2.$$

If your mass is 60 kg, calculate the effective force on you due to you fighting against Beowoof.

EXERCISE 4.3

A batter hits a baseball with a force of $F = (1.30 \times 10^4\,\text{N})\sin(1.05\,\text{ms}^{-1})t$ for 3.00 ms. A baseball is thrown so it has a velocity of 35.0 m/s toward the batter just before it hits the bat. Assuming the baseball has a mass of 0.300 kg, (a) calculate the impulse delivered to the ball over a 3.00 ms time period, and (b) determine the velocity of the baseball just after it leaves the bat.

EXERCISE 4.4

Consider a pinball machine: the pinball is in the contact with the plunger for exactly 0.50 seconds. During that time, the velocity of the ball is given by: $v(t) = (3.0\,\text{m/s}^2)t - (2.0\,\text{m/s}^3)t^2$. If the pinball has mass of 30 g, find the force, $F(t)$, that the plunger exerts on the pinball during the period of contact.

EXERCISE 4.5

Performing a bench press of 80 kg, a man exerts a force upward on the bar of $F(t) =$ $(80\,\text{kg})(9.8\,\text{m/s}^2)\left[1 + \sin(2\pi\,\text{s}^{-1})t\right]$ where $0 < t < 1\,\text{s}$. Calculate the expression for the total momentum of the bar as a function of time. What is the momentum at $t = 0.5\,\text{s}$ and at $t = 1.0\,\text{s}$? [Hint: remember gravity acts on the bar also.]

EXERCISE 4.6

An inexperienced dump truck driver forgets to lower the bed of his dump truck before he begins to drive. The bed was filled with sand when he started his drive, but the sand starts leaking out of the back of the truck so that the mass of the truck with the sand decreases as:

$$m(t) = 3,000\,\text{kg} - (10\,\text{kg/min}^{1/2})t^{1/2} - (3\,\text{kg/min})t.$$

Calculate the force required in order to keep the truck moving at a constant velocity of $20\,\text{m/s}$. (Assume there is no friction in the problem.)

EXAMPLE 4.7

A paint mixing machine shakes a can of paint by applying a force of $F(t) =$ $60\,\text{N}\cos(75\,\text{s}^{-1})t$. If the can of paint has mass of $3.5\,\text{kg}$, find its velocity as a function of time. Find the maximum velocity of the can and the time at which it first occurs.

EXERCISE 4.8

Fire extinguishers are filled with water or foam. There is a fire in the middle of the winter, and a man tries to put out the fire by using the fire extinguisher. He wasn't thinking when he started using the fire extinguisher because he was standing on ice (so there was no friction between his feet and the ground). The mass of the man and the fire extinguisher decreases as the fire extinguisher is used. The mass is given by $m(t) = 90\,\text{kg} - (0.03\,\text{kg/s})t$. When the man starts the fire extinguisher, he begins to slide backwards. His velocity (with the fire extinguisher) is $v = (0.33\,\mathrm{m/s^{3/2}})t^{1/2}$. Calculate the force on the man due to using the fire extinguisher.

EXERCISE 4.9

A bead on a horizontal wire starts at rest and experiences a force given by:

$$F(t) = (0.75\,\mathrm{N}) + (6.75\,\mathrm{N/s^{1/2}})t^{1/2} - (1.3\,\mathrm{N})e^{-(0.20\,\mathrm{s^{-1}})t}.$$

Find an expression for the momentum as a function of time.

EXAMPLE 4.10

A bullet of mass 5.0 g is shot into a large block of wood. If the force on the bullet due to friction with the block of wood is described by: $F = -(4.0\,\frac{\mathrm{N \cdot s}}{\mathrm{m}})v$, and the initial velocity of the bullet is 50 m/s when it enters the block, calculate the velocity of the bullet 3.0 milliseconds after it enters the block.

EXAMPLE 5.1

A body of mass m moves in a horizontal direction such that at time t its position is given by

$$x(t) = \alpha t^4 - \beta t^3 + \gamma t$$

where α, β, γ are constants. (a) What is the acceleration of the body. (b) What is the force acting on the body?

SOLUTION

[Given: body of mass m, moving along the line with position function $x(t) = \alpha t^4 - \beta t^3 + \gamma t$. Find: (a) $a(t)$, (b) $F(t)$.]

The solution to the problem is based on Newton's second law which relates the kinematical property of the motion, acceleration with the force that acts on the body.

(a) Given a position of the body as a function of time the acceleration is obtained by computing the second time derivative of the position,

$$
\begin{aligned}
a(t) &= \frac{d^2 x(t)}{dt^2} = \frac{d^2}{dt^2}\left(\alpha t^4 - \beta t^3 + \gamma t\right)\\
a(t) &= \frac{d}{dt}\left(\frac{d}{dt}\left(\alpha t^4 - \beta t^3 + \gamma t\right)\right) \quad\quad \text{mistake}\\
&= \frac{d}{dt}\left(4\alpha t^3 - 3\beta t^2 + \gamma\right) \quad\quad \text{okay in end}\\
&= 12\alpha t^2 - 6\beta t \ .
\end{aligned}
$$

(b) By Newton's 2nd law of motion, the force that acts on the body is equal to the time derivative of the momentum of the body, $F = dp/dt$, where the momentum or the quantity of motion is expressed as a product of the velocity and the mass, $p = mv$. When the mass of the body is constant, then the time derivative in the force relation only acts on the velocity yielding, $F = mdv/dt = ma$. Hence, the force acting on the body is

$$F = ma = m\left(12\alpha t^2 - 6\beta t\right) = 12m\alpha t^2 - 6m\beta t \ .$$

Notice that in this problem the force is not constant – it changes with time. This will be true in general except when the position is a quadratic function of time.

EXAMPLE 5.2

A body of mass 40 kg is suspended at the lower end of a light vertical chain and is being pulled up vertically, see Fig. EG-5.2.

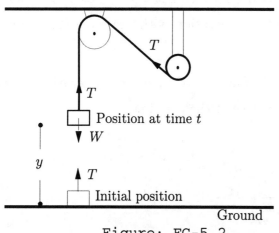

Figure: EG-5.2

Initially the body is at rest and the pull of the chain is 588 N. As the mass rises the pull gets smaller uniformly at a rate of 35 N/m. Find the velocity of the body once it has been raised to a height of 10 m.

SOLUTION

[Given: $m = 40\,\text{kg}$, $v_0 = 0$, $T_0 = 588\,\text{N}$, $R = dT/dy = 35\,\text{N/m}$. Find: v at the height of $y = 10\,\text{m}$.]

Assume that at time t the body is at a height y above the ground. At that moment the total pull on the body is the difference between the initial pull T_0 and the amount by which the initial pull is reduced because the body is elevated (recall that the pull decreases with height). Therefore, $T = T_0 - Ry$. In addition to the pull, the weight, $W = mg$ also acts on the body. By Newton's second law the net force on the body, $T - W = T - mg$, equals to the product of the acceleration and the mass of the body,

$$T - mg = T_0 - Ry - mg = m\frac{d^2y}{dt^2}\ .$$

To find the velocity we need to integrate the equation of motion. Because the net force on the body depends on the height, y, the equation of motion can not be integrated directly. However, multiplying the acceleration by dy and using the relation, $vdt = dy$ yields

$$\frac{d^2y}{dt^2}\,dy = \frac{dv}{dt}\,dy = \frac{dv}{dt}\,v\,dt = v\,dv = \frac{1}{2}d(v^2).$$

Hence, the equation of motion can be rewritten as

$$m\frac{dv}{dt} = \frac{m}{2}\frac{d(v^2)}{dy} = T_0 - mg - Ry\ .$$

This is easy to integrate. Let V be the velocity object at the height $y_{10} = 10\,\text{m}$. Then, (selecting the origin of the coordinates on the ground)

$$\int_0^{y_{10}} \frac{d(v^2)}{dy}\,dy = \int_0^{V^2} d(v^2) = \int_0^{y_{10}} \left(\frac{T_0 - mg}{m/2} - \frac{2R}{m}y\right)\,dy\ .$$

35

The result of the integration is:

$$V^2 = \frac{2(T_0 - mg)}{m}y_{10} - \frac{R}{m}y_{10}^2 \ .$$

Inserting the numbers the square of the velocity evaluates to:

$$V^2 = 8.0 \, \mathrm{m^2/s^2} \ .$$

Hence, the speed is $V = 2.8 \, \mathrm{m/s}$. We still need to find the direction of the velocity. For $0 \leq y \leq 10 \, \mathrm{m}$ the net force, $T_0 - mg - Ry$, is positive at first and becomes negative. Hence, the acceleration is negative at the height of $10 \, \mathrm{m}$. Note that the maximum height that the body can achieve is $10.9 \, \mathrm{m}$. Since the body started from the rest the velocity, on the way up is directed in the upward (positive) direction. If the body has reached the maximum height and is on its way down, the velocity points down and is negative.

EXAMPLE 5.3

A space-ship in a force-free space sweeps up stationary interplanetary debris at a rate $dM/dt = cv$, where c is a constant that depends on the cross-section of the space-ship in a direction of its motion. Assume that all of the space-dust that is encountered by the space-ship along its path sticks to the space-ship's front surface and that the cloud of the space-dust is uniform in its composition. (a) Find the expression for the constant c. (b) Find the deceleration of the space-ship.

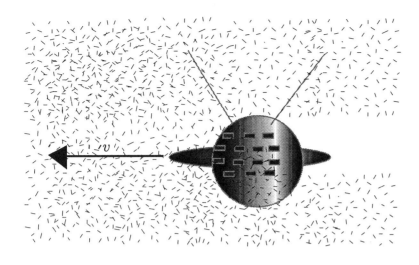

Figure: EG-5.3

SOLUTION

[Given: a satellite in a force-free space filled with stationary debris – space-dust! The dust is collected on the surface of the satellite at a rate $dM/dt = cv$. Find: (a) the constant c; (b) acceleration $a(t)$.]

Let us consider the system consisting of a cloud of space-dust and the space-ship. There are no external forces acting on the system. This means that the net force acting on the system is zero. However, there is a force that acts on the space-ship as it is moving through the cloud of space-debris. That force is equal in magnitude and opposite in direction to the force that the space-ship exerts on the debris (Newton's third law). Alternatively, we can think that the space-ship slows down because of the resistive force that is exerted by the space-dust. As the space-ship moves through the space-dust, the particles of the space-dust that are originally at rest, collide with the space-ship, stick to it and continue moving with it. In other words, some of the space-ship's momentum is transferred to the particles of the space-dust. This momentum loss by the space-ship is the origin of the resistive force with which the space dust acts on the space-ship.

(a) Let us first calculate the rate at which the space-dust is collected on the space-ship. Note that while the space-ship is moving through the cloud of space-debris its mass increases because the particles of the dust stick to it. Hence, the mass of the space-ship is a function of time. It has been assumed that the rate at which the space-ship acquires mass is proportional to the speed of the space ship, that is $dM/dt = cv$. This implies that the mass is acquired at the rate which is proportional to the rate at which the volume is swept by the space-ship. To see that, let the speed of the space-ship at time t be v. Then, during the short time interval, Δt, following the time instant, t, the space-ship moves a distance, $\Delta x = v\Delta t$. Considering the time interval Δt which is very short (infinitesimal), we can assume that the velocity of the space-ship will not change much and in the limit $\Delta t \to 0$ the result will be exact. Because the space-dust is stationary prior to the collision and all of the dust particles that undergo a collision stick to the surface of the space-ship, the increase of the space-ship's mass is equal to the mass of the space debris within the volume swept by the space-ship during time interval Δt. The total volume that is swept by the space-ship during the time interval Δt equals to the product of the distance $v\Delta t$ that the space-ship traveled and the cross-sectional area A of the space-ship. Hence, $\Delta V = A\Delta x = vA\Delta t$. Since the cloud of space dust is uniform, the total mass within the volume ΔV equals to the product of the density ρ of the space dust and the volume within which it is contained, $\Delta m = \rho\Delta V = \rho vA\Delta t$. Hence, the amount of mass per unit time acquired by the space-ship is $\Delta m/\Delta t = \rho Av$. Taking time Δt to be infinitesimal, yields $dM/dt = \rho Av$, where A is the cross-sectional area of the space-ship, ρ is the density of the dust cloud and v is the speed of the space-ship at time instant t. Comparing this with the relation assumed in the problem, $dM/dt = cv$, we find the expression for the proportionality constant, $c = \rho A$.

(b) Next, we calculate the deceleration of the space-ship. The deceleration is obtained by finding the force on the space-ship while it is moving. The net force on the space-ship – dust-cloud system is zero:

$$F_{net} = 0 = \frac{d}{dt}(Mv) = \frac{dM}{dt}v + M\frac{dv}{dt} .$$

The second term in the equation is the force on the space-ship. Hence,

$$F_{space\ ship} = M\frac{dv}{dt} = -v\frac{dM}{dt} = -cv^2 .$$

Dividing the force that acts on the space-ship by the mass M of the space-ship gives the deceleration (negative acceleration):

$$a = -\frac{c}{M}v^2 .$$

Note that this is not a simple relationship because the mass M of the space-ship changes with time.

EXERCISES

EXERCISE 5.1

A body of mass m moves in a vertical direction such that at time $t > 0$ its position is given by

$$y(t) = at^{3/2} - bt + c$$

where a, b, c are constants. (a) Calculate the acceleration of the body. (b) What is the force acting on the body?

39

EXERCISE 5.2

Consider a bead having mass m that is free to move on a thin fixed wire bent into a horizontal circle of radius R, see Fig. EX-5.2 As the bead moves along the perimeter of the circle it experiences a friction force that slows it down. Assuming that the bead is given the initial speed v_0 and that the coefficient of kinetic friction is μ_k find the speed of the bead as a function of time.

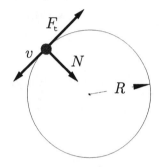

Figure: EX-5.2

EXERCISE 5.3

A space vehicle has a mass of $200\,\text{kg}$ and cross-sectional area of $2 \times 10^4\,\text{cm}^2$. It travels in a region of space without appreciable gravitational force through a rarefied atmosphere with mass density of $2 \times 10^{-15}\,\text{g/cm}^3$. Let the initial speed of the space vehicle be $7.6 \times 10^5\,\text{cm/s}$ (these are the approximate values for the satellite at the altitude of $500\,\text{km}$ above the surface of the Earth). (a) Assuming that the particles of the atmosphere stick to the vehicle as it moves through it at a rate, $dM/dt = cv$, find the numerical value of the constant c. (b) Using the conservation of momentum, $Mv = M_0 v_0$, show that the space vehicles deceleration is given by $dv/dt = -cv^3/M_0 v_0$. (c) Find how long it will take for the vehicle to have its speed reduced to 90% of the initial speed.

EXERCISE 5.4

A space vehicle ejects fuel backwards at the velocity $\mathbf{V_0}$ relative to the vehicle. As a result, the space vehicle is loosing mass. Let us assume that the mass is lost a constant rate $dM/dt = -\alpha$. Neglecting gravity, find the speed of the vehicle as a function of time.

EXERCISE 5.5

A rocket of initial mass M_0 burns an adjustable amount of fuel β (measured in kg/s) and climbs up from the launch pad on the surface of the Earth. The fuel is ejected straight down with velocity $\mathbf{V_0}$ relative to the rocket. (a) Find the rate at which the fuel has to be burned so that the rocket remains stationary short distance above the ground. (b) Assuming that the amount of fuel used per second remains constant (but greater than the rate found in part (a)) find the upward velocity of the rocket as a function of time. (c) Assuming that the fuel leaves the rocket at the speed $V_0 = 1.62 \times 10^3$ m/s (five times grater than the speed of sound), find the rocket speed when the rocket has a total mass equal to $3M_0/4$. Assume that $\alpha = 2M_0 g / V_0$.

EXERCISE 5.6

Consider a flexible chain or a rope that is suspended above a stationary platform with one end initially just touching the platform. Then release the chain and allow it to freely fall onto the platform. Find the force that the platform experiences while the chain is falling onto it.

EXERCISE 5.7

A flexible massless rope is placed over a cylinder of radius R. A tension T is applied at each end of the rope which remains stationary. Show that each small segment $R\Delta\theta$ of the rope, see figure EX-5.7 which is in contact with the cylinder pushes against the cylinder with force $T\Delta\theta$ in the radial direction. By integrating the force exerted on the cylinder by all small segment that are in contact with the cylinder, show that the net vertical force on the cylinder is $2T$ and that the net horizontal force is zero. [HINT: assume that $\Delta\theta$ is infinitesimal.]

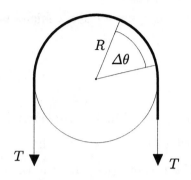

Figure: EX-5.7

EXERCISE 5.8

A horizontal force $F = A + Bt^3$ acts on a 2 kg object where $A = 5.0\,\text{N}$ and $B = 2.0\,\text{N/s}^2$. What is the horizontal velocity of the object 4 seconds after it has started from rest?

EXERCISE 5.9

A small pellet of mass m falls from rest through a medium that exerts a resisting force that varies directly with the square of the speed, $F_r = mcv^2$. Find the speed of the pellet as function of time. What is the terminal velocity? Howl long it will take before the object reaches the terminal velocity?

EXERCISE 5.10

The end of a chain, of mass per unit length λ at rest on a table top at $t = 0$, is lifted vertically at constant speed v. Evaluate the upward force as a function of time.

EXAMPLE 7.1

A point particle of mass m is distance l away from an infinite massive plane with uniform surface mass density σ. Calculate the gravitational force with which the plane acts on the point mass.

SOLUTION

[Given: a point particle of mass m, distance l away from an infinite massive plane with uniform surface mass density σ. Find: **F**.]

To find the gravitational force with which an infinite massive plane acts on the point mass, begin by setting up the geometry so that the calculation is as easy as possible. A convenient choice of the coordinate axis and the corresponding position vectors are given in Fig. EG-7.1.

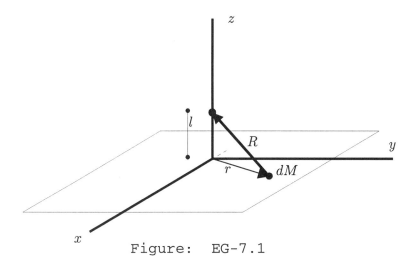

Figure: EG-7.1

It is practical to choose the coordinate axis such that the massive plane coincides with x-y plane of the coordinate system. The point mass is placed on the z-axis, distance l away from the plane. The position vector of the point mass is $\mathbf{r}_m = l\hat{z}$. To calculate the force imagine that the infinite massive plane is divided into small segments each having mass $dM = \sigma dx dy$. A location of a small mass segment dM is given by the position vector $\mathbf{r} = x\hat{x} + y\hat{y}$. The vector which points from the mass segment dM of the plane toward the point mass m is then given by $\mathbf{R} = \mathbf{r}_m - \mathbf{r} = -x\hat{x} - y\hat{y} + l\hat{z}$. Using Newton's gravitational force law, the expression for the force with which the segment of mass dM acts on the point mass m is,

$$d\mathbf{F} = -GmdM\frac{\mathbf{R}}{R^3} = -Gm\sigma dx dy \frac{-x\hat{x} - y\hat{y} + l\hat{z}}{(x^2 + y^2 + l^2)^{3/2}} \; .$$

The total force is obtained by integrating the force $d\mathbf{F}$ over the massive plane,

$$\mathbf{F} = \int\limits_{PLANE} d\mathbf{F} \; .$$

To proceed, it is best to consider each component of the force separately. Let us first consider the x-component.

$$F_x = +Gm\sigma \int\limits_{-\infty}^{\infty} dx \int\limits_{-\infty}^{\infty} dy \; \frac{x}{(x^2 + y^2 + l^2)^{3/2}} \; .$$

It is easy to see that this integral is zero. The symmetry argument is the easiest one to understand. Note that the integration over the variable x is over a symmetric interval extending from $-\infty$ to $+\infty$. Also note that the integrand is an odd function of x (it changes a sign when x changes the sign). The form of an integral over the x variable is, $I_x = \int_{-a}^{a} f(x)dx$ where $f(x) = -f(-x)$. Then, $I_x = \int_{-a}^{0} f(x)dx + \int_{0}^{a} f(x)dx$. In the first term we can make a change of variable, $x = -t$, which yields $\int_{-a}^{0} df(x)dx = \int_{a}^{0} f(-t)(-dt) = (-1)^2 \int_{0}^{a}(-f(t))dt = -\int_{0}^{a} f(x)dx$, where in the last step we have renamed an integration variable back to x. This implies $I_x = -\int_{0}^{a} f(x)dx + \int_{0}^{a} f(x)dx = 0$. Hence, $F_x = 0$. Using the same argument we can show that the y-component also vanishes – try it.

Only the z-component of the gravitational force is nonzero and reads

$$F_z = -Gml\sigma \int\limits_{-\infty}^{\infty} dx \int\limits_{-\infty}^{\infty} dy \; \frac{1}{(x^2 + y^2 + l^2)^{3/2}} \cdot$$

This integral can be evaluated in Cartesian coordinates or in polar coordinates in the plane. Here, we will use Cartesian coordinates. You should try to do it in planar polar coordinates!* To evaluate this double integral, first do the integration over one of the variables while keeping the other variable fixed. Then you are left with a single integral. Let us first integrate over the y-variable and keep x-variable fixed. For the ease of notation, let us introduce a quantity $c^2 = x^2 + l^2$. It is a good idea to perform a change of variables because we can simplify the y integral. A convenient new integration variable, call it t, is defined by a transformation $y = c \sinh t$. Since the range of the y variable is $-\infty \le y \le +\infty$, the new variable t has the same range, $-\infty \le t \le +\infty$. (Consult your algebra and calculus texts if you need to remind yourselves of the properties of the hyperbolic sine function!) Therefore, $dy = c \cosh t\,dt$. The denominator of the integrand becomes: $(x^2 + y^2 + l^2)^{3/2} = (c^2 + y^2)^{3/2} = (c^2)^{3/2}(1 + \sinh^2 t)^{3/2} = c^3(\cosh^2 t)^{3/2} = c^3 \cosh^3 t$, where the property $\cosh^2 t - \sinh^2 t = 1$, of the hyperbolic sine and hyperbolic cosine functions was used. The integral over the y variable is of the form

$$\int\limits_{-\infty}^{\infty} \frac{c \cosh t}{c^3 \cosh^3 t}\,dt = \frac{1}{c^2} \int\limits_{-\infty}^{\infty} \frac{dt}{\cosh^2 t}.$$

Note that the dependence on the x-variable is hidden in the parameter c. To evaluate the integral over the variable t recall the following formula from the differential

---

* To evaluate the integral in planar polar coordinates make a transformation of integration variables, $x = r \cos \phi$, $y = r \sin \phi$, where $r = \sqrt{x^2 + y^2}$ and $\tan \phi = y/x$. Then, $dxdy = rdrd\phi$ (consult your calculus text for details). The range of new variables is, $0 \le r \le \infty$, and $0 \le \phi \le 2\pi$.

calculus: $\frac{d}{dt}\tanh t = \frac{1}{\cosh^2 t}$. Hence,

$$\frac{1}{c^2}\int_{-\infty}^{\infty} d\left(\tanh t\right) = \frac{1}{c^2}[\tanh t]_{-\infty}^{\infty} = \frac{1}{c^2}[\tanh \infty - \tanh(-\infty)] = \frac{1}{c^2}[1-(-1)] = \frac{2}{c^2}.$$

At this point the expression for the z-component of the force has the form:

$$F_z = -2Gm\sigma l \int_{-\infty}^{\infty} \frac{dx}{x^2 + l^2} \, .$$

Again, the x integral is evaluated by performing the variable transformation. The following substitution works: $x = l\tan t$. Then, $l^2 + x^2 = l^2(1+\tan^2 t) = l^2(1 + \sin^2 t/\cos^2 t) = l^2(\cos^2 t + \sin^2 t)/\cos^2 t = l^2/\cos^2 t$. To obtain the expression after the last equal sign the trigonometric identity, $\cos^2 t + \sin^2 = 1$ was used. The differential element becomes, $dx = ldt/\cos^2 t$. The region of integration in the new variable is $-\frac{\pi}{2} \le t \le \frac{\pi}{2}$, (recall, $\tan \frac{\pi}{2} = +\infty$). Therefore, the integral takes the form and evaluates to:

$$F_z = -\frac{2Gm\sigma d^2}{d^2} \int_{-\frac{\pi}{2}}^{\frac{\pi}{2}} \frac{dt}{\cos^2 t}\cos^2 t = -Gm\sigma \int_{-\frac{\pi}{2}}^{\frac{\pi}{2}} dt = -2\pi Gm\sigma \, .$$

The force with which the infinite massive plane attracts a point mass is directed straight from the point mass toward the massive plane and has magnitude $F = 2\pi Gm\sigma$. Note that the force is constant.

EXAMPLE 7.2

Imagine an infinitely long, very thin, uniform, massive rod with mass density per unit length λ and a point mass M placed a distance l away from the rod. Calculate the attractive gravitational force with which the rod acts on the point mass.

SOLUTION

[Given: infinitely long, very thin, massive rod with uniform mass density per unit length λ. A point mass M at distance l from the rod. Find: $\mathbf{F}$.]

The geometry of the problem is given in Fig. EG-7.2.

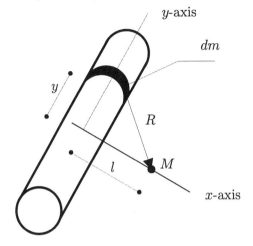

Figure: EG-7.2

To calculate the force with which the rod acts on the point mass imagine dividing the rod into small (infinitesimal) segments of mass $dm = \lambda dy$ and use Newton's gravity law to write the force that this segment exert on the point mass. The position vector of the mass segment dm is $y\hat{y}$ and that of the point mass M is $l\hat{x}$. The vector $\mathbf{R} = l\hat{x} - y\hat{y}$ points from the mass segment dm toward the point mass M. Hence, the gravitational force between the mass segment dm and the point mass M is:

$$d\mathbf{F} = -GM\lambda dy \frac{l\hat{x} - y\hat{y}}{(l^2 + y^2)^{3/2}} \ .$$

The total force is found by integrating over the length of the rod:

$$\mathbf{F} = \int_{ROD} d\mathbf{F} = -GM\lambda \int_{-\infty}^{\infty} dy \frac{l\hat{x} - y\hat{y}}{(l^2 + y^2)^{3/2}} \ .$$

47

To proceed, it is best to consider two components of the force separately. The y-component of the force is given by

$$F_y = GM\lambda \int\limits_{-\infty}^{\infty} \frac{y \, dy}{(l^2 + y^2)^{3/2}} \ .$$

Using the symmetry argument (see Example problem 7.1 for details) it is easy to see that the integral is zero; $F_y = 0$. Hence, the force is directed straight toward the rod:

$$\mathbf{F} = \hat{x} F_x = -GMl\lambda\hat{x} \int\limits_{-\infty}^{\infty} \frac{dy}{(l^2 + y^2)^{3/2}} \ .$$

The integral that needs to be evaluated is identical to the one worked out in the Example problem 7.1. Make a substitution, $y = l \sinh t$, and follow the steps outlined in the Example problem 7.1. The integral becomes

$$F_x = -\frac{GM\lambda}{l} \int\limits_{-\infty}^{\infty} \frac{dt}{\cosh^2 t} \ .$$

Since, $\int_{-\infty}^{\infty} dt/\cosh^2 t = 2$, the force is

$$\mathbf{F} = -\frac{2Gm\lambda}{l}\hat{x} \ .$$

The force is directed toward the rod. You may also do the calculation in cylindrical coordinates. Try it!

EXAMPLE 7.3

Imagine a point particle of mass m placed a distance r away from an infinitely long, thin, massive rod with uniform mass density λ The particle has no initial velocity and it is attracted by the gravitational force toward the rod. What will be the velocity of the particle when it is distance r_0 away from the rod.

SOLUTION

[Given: particle of mass m initially at rest, $v_0 = 0$; thin, uniform, infinitely long rod with mass density per unit length λ acts with gravitational force on the particle. Find: v.]

The gravitational force with which the rod acts on the particle is directed toward the rod and is given by (see Example problem 7.2)

$$F = -\frac{2Gm\lambda}{r} .$$

The direction pointing away from the rod is chosen as a positive direction. In writing the force equation we took into the account that the motion is along the straight line. This means that to keep track of the directions we can work with the scalar value of the force. Applying Newton's second law the equation of motion for the point mass reads:

$$m\frac{dv}{dt} = -\frac{2Gm\lambda}{r} .$$

To find the scalar value of the velocity, the acceleration dv/dt is integrated over time. However, since the acceleration is not given as an explicit function of time, we must first transform the equation of motion into a form more suitable for integration. Hence, multiply the equation by $vdt = dr$. This yields,

$$vdt\frac{dv}{dt} = vdv = -\frac{2Gm\lambda}{r}dr .$$

The expression after the first equal sign becomes, $vdv = d(v^2/2)$. Since $dr/r =$

49

$d(\ln r)$, the equation becomes easy to integrate:

$$\frac{1}{2} \int_0^v d(v^2) = \frac{v^2}{2} = -2Gm\lambda \int_r^{r_0} d(\ln r) = -2Gm\lambda \ln \frac{r_0}{r} \; .$$

Also, since $r > r_0$ the logarithm as it is written is negative; together with an overall negative sign the expression is positive! Taking the square root yields

$$v = -\sqrt{2G\lambda \ln \frac{r}{r_0}} \; .$$

We took the negative sign of the square root because the particle is moving toward the rod – negative direction by our conventions)

Note that if we wish to find a trajectory as a function of time we have to calculate a rather difficult integral of the form $\int \frac{dr}{\sqrt{\ln r/r_0}}$!

EXERCISES

EXERCISE 7.1

Imagine two, infinite in size, massive planes with uniform mass densities σ_1 and σ_2 respectively, placed parallel to each other and separated by distance l. It is plausible that the total gravitational attraction between these two planes is infinite. However, the force per unit area is finite. Calculate the force per unit area between the two planes and show that it is finite. Also show that the total force is infinite.

EXERCISE 7.2

Imagine two, infinite in size, very thin, massive rods with uniform mass densities λ_1 and λ_2 respectively, placed parallel to each other and separated by distance l. It is plausible that the total gravitational attraction between these two rods is infinite. However, the force per unit length is finite. Calculate the force per unit length between the two rods and show that it is finite. Also show that the total force is infinite.

EXERCISE 7.3

Consider a particle, initially a distance r_0 away from an infinitely long, very thin massive rod with uniform mass density per unit length λ. The particle is launched in a direction straight away from the rod with some velocity v_0. What must be value of the launch velocity if the particle is to escape the gravitational attraction of the rod? Can it escape? If not, what is the maximal distance form the rod that the particle can reach?

EXERCISE 7.4

Consider a massive, very thin ring of radius R and mass M. A particle of mass m is located along the axis of the ring distance l away from its center in a direction perpendicular to the plane of the ring. What is the gravitational force with which the ring acts on the particle?

EXERCISE 7.5

Imagine a particle of mass m placed, distance r above a massive ring of radius R and mass M just like in Exercise 7.4. The particle starts moving from the rest toward the ring. What will be the velocity of the particle when it reaches the center of the ring. How far will the particle go on the other side of the plane of the ring?

EXERCISE 7.6

Imagine two uniform, massive spheres of masses M_1 and M_2 and radii R_1 and R_2 respectively, separated by distance (center to center) l ($l > R_1 + R_2$). Show that the gravitational force between the two spheres is given by Newton's law!

EXERCISE 7.7

A particle of mass m can move in a circular orbit of radius R around the source of the gravitational force. Show, by direct calculation that the circular orbit satisfies the equation of motion of the particle. [HINT: consider an equation of the circle parameterized in some suitable way. It helps to choose the coordinate system carefully.]

EXERCISE 7.8

A satellite of mass $m = 7000\,\text{kg}$ has been placed into a circular orbit around the Sun having the same radius as the orbit of the Earth, $R = 1.50 \times 10^{11}\,\text{m}$ with a goal of sending it on a trip outside the Solar system. What minimal velocity must the satellite be given so that it can travel outside the solar system? The mass of the Sun is $M_S = 1.987 \times 10^{30}\,\text{kg}$.

EXERCISE 7.9

A space vehicle is to be launched from the surface of the Earth toward the Moon, If the vehicle is to make it exactly to the point where the gravitational pull of the Earth and the Moon cancel each other, what velocity must the vehicle be given at launch? [$M_E = 5.975 \times 10^{24}\,\text{kg}$, $R_E = 6.371 \times 10^6\,\text{m}$, $M_M = 7.35 \times 10^{22}\,\text{kg}$, $R_M = 1.74 \times 10^6\,\text{m}$ $R_{EM} = 3.85 \times 10^8\,\text{m}$.]

EXAMPLE 8.1

A child on a tricycle rides along a straight, horizontal sidewalk, so that her position varies as: $l(t) = (7.00\,\text{cm/s})t - (3.00\,\text{cm/s}^{1/2})t^{1/2}$, $(t > 1.00\,\text{s})$. Assuming that the radius of the tire is 14.0 cm, calculate the angular acceleration of the front tire as a function of time.

SOLUTION

[Given: $l(t) = At - Bt^{1/2}$, where $A = 7.00\,\text{cm/s}$ and $B = 3.00\,\text{cm/s}^{1/2}$, $r = 14.0\,\text{cm} = 0.140\,\text{m}$. Find: $\alpha(t)$.]

This problem investigates the relationship between linear and angular quantities. From the chapter, we know $l = r\theta$, so we can find the angular displacement from the child's linear displacement:

$$\theta = \frac{l}{r} = \frac{A}{r}t - \frac{B}{r}t^{1/2} = \frac{7.00\,\text{cm/s}}{14.0\,\text{cm}}t - \frac{3.00\,\text{cm/s}^{1/2}}{14.0\,\text{cm}}t^{1/2}$$
$$=(0.500\,\text{rad/s})t - (0.214\,\text{rad/s}^{1/2})t^{1/2}$$
$$\theta(t) = Ct - Dt^{1/2},$$

where $C = 0.500\,\text{rad/s}$ and $D = 0.214\,\text{rad/s}^{1/2}$.

Now we can use the relationships between θ, ω, and α to get to α. $\omega = \frac{d\theta}{dt}$ and $\alpha = \frac{d\omega}{dt}$, so

$$\omega = \frac{d\theta}{dt} = \frac{d}{dt}\left(Ct - Dt^{1/2}\right) = C - \tfrac{1}{2}Dt^{-1/2},$$
$$\alpha = \frac{d\omega}{dt} = \frac{d}{dt}\left(C - \tfrac{1}{2}Dt^{-1/2}\right) = 0 + \tfrac{1}{4}Dt^{-3/2}.$$

So, substituting for D, $\alpha(t) = \tfrac{1}{4}(0.214\,\text{rad/s}^{1/2})t^{-3/2}$, or $\alpha = (0.0536\,\text{rad/s}^{1/2})t^{-3/2}$. Notice that α does indeed have units of rad/s^2 as it should.

EXAMPLE 8.2

Meadowlark Lemon begins to spin a basketball on his finger. He flicks it up onto his finger so that it has an initial angular velocity of 1.0 rotations per second. He keeps pushing on it to keep it speeding up. During this time, he exerts a constant torque of 0.067 N·m on the ball for three seconds. Calculate the angular displacement as a function of time for the first three seconds of rotation. Assume the basketball has a mass of 0.35 kg, evenly distributed at radius of 14 cm.

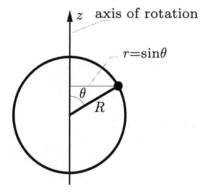

Figure: EG-8.2

SOLUTION

[Given: $\omega_0 = 1.0\,\text{rot/sec} = 2\pi\,\text{rad/s}$, $\tau = 0.067\,\text{N·m}$, $t = 3.0\,\text{s}$, for basketball: $m = 0.35\,\text{kg}$, $R = 14\,\text{cm} = 0.14\,\text{m}$. Find: $\theta(t)$ for first 3 seconds.]

This is a tricky problem, but it should prove instructional. First, we know that $\tau = I\alpha$, where I is the moment of inertia of a thin spherical shell and α is the angular acceleration. So, let's derive the expression for the moment of inertia of a thin spherical shell. We know $I = \int r^2\,dm$. So, we need to determine what dm is. In three-dimensions $dm = \rho dV$, in two dimensions $dm = \rho_A\,dA$, where ρ_A is the areal density. This problem is 2-dimensional, do $dm = \rho_A\,dA$. Since the mass is evenly distributed $\rho_A = \frac{m}{A} = \frac{m}{4\pi r^2}$. And $dA = R^2 \sin\theta\,d\theta\,d\phi$, (see your calculus

54

book). So, $I = \int r^2 dm$ or

$$I = \int r^2 \left(\frac{m}{4\pi R^2}\right) R^2 \sin\theta \, d\theta \, d\phi$$

$$= \frac{m}{4\pi} \int r^2 \sin\theta \, d\theta \, d\phi$$

We do not integrate over dr because we have a thin shell. ($R = 14\,\text{cm}$). But as θ changes, the distance away from the axis of rotation changes: So, $r = R\sin\theta$, where R is the radius of the ball. So,

$$I = \frac{m}{4\pi} \int\limits_0^{2\pi} \int\limits_0^{\pi} (R\sin\theta)^2 \sin\theta \, d\theta \, d\phi$$

$$= \frac{mR^2}{4\pi} \left(\int\limits_0^{\pi} \sin^3\theta \, d\theta\right) \phi\Big|_0^{2\pi} = \frac{mR^2}{4\pi} \int\limits_0^{\pi} \sin^3\theta \, d\theta (2\pi)$$

$$= \frac{mR^2}{4\pi} (2\pi) \int\limits_0^{\pi} \sin^3\theta \, d\theta \quad \text{recall: } \sin^2\theta = 1 - \cos^2\theta$$

$$= \frac{mR^2}{2} \int\limits_0^{\pi} (1 - \cos^2\theta) \sin\theta \, d\theta$$

now, let $x = \cos\theta$, $dx = -\sin\theta \, d\theta$ (limits are now $+1$ to -1)

$$= \frac{mR^2}{2} \int\limits_{+1}^{-1} -(1 - x^2) \, dx = \frac{mR^2}{2} \int\limits_{-1}^{+1} (1 - x^2) \, dx = \frac{mR^2}{2} \left(x - \tfrac{1}{3}x^3\right) \Big|_{x=-1}^{x=+1}$$

$$= \frac{mR^2}{2} \left[(1 - \tfrac{1}{3}) - (-1 + \tfrac{1}{3})\right] = \frac{mR^2}{2} \left[\tfrac{2}{3} - (-\tfrac{2}{3})\right] = \frac{mR^2}{2} \left(\tfrac{4}{3}\right) = \tfrac{2}{3}mR^2$$

So, I (for a spherical shell) is $\tfrac{2}{3}mR^2$ (see the table in the text).

$$\Rightarrow I = \tfrac{2}{3}(0.35\,\text{kg})(0.14\,\text{m})^2 = 4.57 \times 10^{-3}\,\text{kg·m}^2$$

(Note: If you want to calculate I for a solid sphere $dm = \rho dV$ where $\rho = \frac{m}{4/3\pi R^3}$ and $dV = r^2 \sin\theta \, dr \, d\theta \, d\phi$ with bound of $0 < r < R$, $0 < \theta < \pi$, and $0 < \phi < 2\pi$).

Now that we calculated I, we can calculate α: $\tau = I\alpha$

$$\alpha = \frac{\tau}{I} = \frac{0.067\,\text{N·m}}{4.57 \times 10^{-3}\,\text{kg·m}^2} = 14.7\frac{(\text{kg·m}/\text{s}^2)m}{\text{kg·m}^2} = 14.7\,\text{s}^{-2} = 14.7\,\text{rad/s}^2$$

Next, we need to use the relations between α, ω, and θ. Remember $\alpha = \frac{d\omega}{dt}$ or $\int_{\omega_0}^{\omega_f} d\omega = \int_{t_0}^{t_f} \alpha\,dt$ and $\omega = \frac{d\theta}{dt}$ or $\int_{\theta_0}^{\theta_f} d\theta = \int_{t_0}^{t_f} \omega\,dt$. So, since we know α, we can find ω:

$$\int_{\omega_0}^{\omega} d\omega = \int_0^t \alpha\,dt \Rightarrow \omega - \omega_0 = \alpha t \Big|_0^t = \alpha t$$

so, $\omega = \omega_0 + \alpha t$ and $\omega_0 = 1\frac{\text{rot}}{\text{s}} \cdot \frac{2\pi\text{rad}}{\text{rot}} = 2\pi\,\text{rad/s}$

Now, we can keep going to find θ:

$$\int_{\theta_0}^{\theta} d\theta = \int_0^t \omega\,dt = \int_0^t (\omega_0 + \alpha t)\,dt \Rightarrow \theta - \theta_0 = \omega_0 t + \tfrac{1}{2}\alpha t^2 \Big|_0^t$$

$$= \omega_0 t + \frac{1}{2}\alpha t^2$$

Since we want θ relative to its starting position, $\theta_0 = 0$, and

$$\theta(t) = \omega_0 t + \frac{1}{2}\alpha t^2 = (2\pi\,\text{rad/s})t + \tfrac{1}{2}(14.7\,\text{rad/s}^2)t^2$$

$$= (6.28\,\text{rad/s})t + (7.33\,\text{rad/s}^2)t^2$$

So, the angular displacement as a function of time for the first three seconds of rotation goes as:

$$\theta(t) = (6.3\,\text{rad/s})t + (7.3\,\text{rad/s}^2)t^2$$

EXAMPLE 8.3

Calculate the moment of inertia of an empty soda can about the center of its radius. The can has a 6.00 cm inner diameter and is 12.0 cm tall. Assume the aluminum is equally distributed around the can, the can in 1.00 mm in thickness, and the density of aluminum is 2.70 g/cm³. (Don't forget that there is a top and a bottom to the can. You may model them as flat disks.)

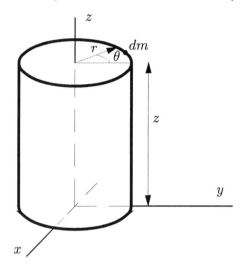

Figure: EG-8.3

SOLUTION

[Given: $R_{\text{can}}(\text{inner}) = 3.00\,\text{cm}$, $h_{\text{can}} = 12.0\,\text{cm}$, $t_{\text{can}} = 1.00\,\text{mm}$, $\rho_{\text{Al}} = 2.70\,\text{g/cm}^3$. Find: I.]

Remember the equation for moment of inertia is $I = \int r^2\,dm$. Also, the moments of inertia add, so we can split the problem into parts.

$$I_{\text{total}} = I_{\text{cylinder}} + I_{\text{top}} + I_{\text{bottom}}\,.$$

The cylinder is 12.0 cm high, 3.00 cm in radius and 1.00 mm thick. The top and the bottom are both solid disks with height 1.00 mm, and radius 3.10 cm. So, we can calculate each of the parts separately and then add them at the end.

First, let's calculate the moment of inertia for the cylinder. Since $dm = \rho \, dV$ and we know ρ for aluminum, we need to get an expression for dV. $dV = r \, dr \, d\theta \, dz$ (from your calculus book), see the diagram for the definitions of r, θ and z.

The limits for the integration are: $3.00\,\text{cm} < r < 3.10\,\text{cm}$, $0 < z < 12.0\,\text{cm}$, and $0 < \theta < 2\pi$. Notice that this is *not* a two-dimensional problem like Example #2, because the shell has a thickness of $0.10\,\text{cm}$. So, $I_{\text{cylinder}} = \int r^2 \, dm = \int_0^{12.0\,\text{cm}} \int_0^{2\pi} \int_{3.00\,\text{cm}}^{3.10\,\text{cm}} r^2 \, \rho r \, dr \, d\theta \, dz$

$$I_{\text{cylinder}} = \rho_{\text{Al}} \int_0^{12.0\,\text{cm}} \int_0^{2\pi} \int_{3.00\,\text{cm}}^{3.10\,\text{cm}} r^3 \, dr \, d\theta \, dz = \rho_{\text{Al}} \left(\int_{3.00\,\text{cm}}^{3.10\,\text{cm}} r^3 \, dr \right) \left(\theta \Big|_0^{2\pi} \right) \left(z \Big|_0^{12.0\,\text{cm}} \right)$$

$$= \rho_{\text{Al}} (2\pi)(12.0\,\text{cm}) \int_{3.00\,\text{cm}}^{3.10\,\text{cm}} r^3 \, dr = \rho_{\text{Al}} (12.0\,\text{cm}) \left(\tfrac{1}{4} r^4 \Big|_{3.00\,\text{cm}}^{3.10\,\text{cm}} \right)$$

$$= \rho_{\text{Al}} \left(\tfrac{\pi}{2} \right) (12.0\,\text{cm})[(3.10\,\text{cm})^4 - (3.00\,\text{cm})^4] = \rho_{\text{Al}} \left(\tfrac{\pi}{2} \right)(12.0\,\text{cm})(11.4\,\text{cm}^4)$$

$$I_{\text{cylinder}} = (2.70\,\text{g/cm}^3)(214\,\text{cm}^5) = 577.8\,\text{g·cm}^2$$

Now, we need to do the top and the bottom. They are the same, so $I_{\text{top}} = I_{\text{bottom}}$. Again, this is like a cylinder, but the bounds of integration are different again: $0 < z < 0.100\,\text{cm}$, $0 < r < 3.10\,\text{cm}$ ($3.10\,\text{cm}$ because the can's full radius with the aluminum is $3.00\,\text{cm}+$ thickness of aluminum walls), and $0 < \theta < 2\pi$. So,

$$I_{\text{top}} = I_{\text{bottom}} = \int r^2 \, dr = \int_0^{0.100\,\text{cm}} \int_0^{2\pi} \int_0^{3.10\,\text{cm}} r^2 \rho_{\text{Al}} r \, dr \, d\theta \, dz$$

$$= \rho_{\text{Al}} \int_0^{3.10\,\text{cm}} r^3 \, dr \left(\theta \Big|_0^{2\pi} \right) \left(z \Big|_0^{0.100\,\text{cm}} \right) = \rho_{\text{Al}} (2\pi)(0.100\,\text{cm}) \int_0^{3.10\,\text{cm}} r^3 \, dr$$

$$= \rho_{\text{Al}} (2\pi)(0.100\,\text{cm}) \left(\tfrac{1}{4} r^4 \right) \Big|_0^{3.10\,\text{cm}} = (2.70\,\text{g/cm}^3) \left(\tfrac{\pi}{2} \right) (9.24\,\text{cm}^5)$$

$$= 39.2\,\text{g·cm}^2 \ .$$

Now we can add them together:

$$I = I_{\text{cylinder}} + I_{\text{top}} + I_{\text{bottom}}$$

$$= 577.8\,\text{g·cm}^2 + 2(39.2\,\text{g·cm}^2)$$

$$I = 6.56 \times 10^2\,\text{g·cm}^2 .$$

and converting to SI units:

$$I = (6.56 \times 10^2\,\text{g·cm}^2)\left(\frac{1\,\text{kg}}{1000\,\text{g}}\right)\left(\frac{1\,\text{m}}{100\,\text{cm}}\right)^2 = 6.56 \times 10^{-5}\,\text{kg·m}^2$$

So, $I_{\text{total}} = 6.56 \times 10^{-5}\,\text{kg·m}^2$. (Remember, when the system is complex, split it up into parts that you know how to calculate.)

EXAMPLE 8.4

A small ball having mass m is attached to an elastic spring with spring constant $k = m\omega^2$. It starts from the origin with speed v_0 and moves along a straight line. Find the trajectory of the particle.

SOLUTION

[Given: $ma = -kx = -m\omega^2 x$; $x(0) = 0$; $v(0) = v_0$. Find: $x(t)$.]

This is a harmonic oscillator problem. The position of a particle as a function of time is found by integrating the equation of motion of the system. The equation of motion for a harmonic oscillator is:

$$a = \frac{d^2 x}{dt^2} = \frac{dv}{dt} = -\omega^2 x .$$

This problem looks straightforward but it isn't because the acceleration is not given as a function of time. To solve the problem, we must multiply the equation of motion by $dx = vdt$. This gives: $adx = -\omega^2 x dx = (dv/dt)vdt = vdv$. The resulting equation is easy to integrate.

$$-\omega^2 \int x dx = -\frac{1}{2}\omega^2 x^2 = \int v dv = \frac{1}{2}v^2 - c \,,$$

where "$-c$" is some integration constant that is determined by the initial conditions. We leave this for the end.

Next, we can rearrange the equation just derived to express the velocity as a function of the position:

$$v = \frac{dx}{dt} = \sqrt{c - \omega^2 x^2} \,.$$

Again, this equation is not ready for immediate integration. To bring the equation in the form which can be integrated, write:

$$dt = \frac{dx}{v(x)} = \frac{dx}{\sqrt{c - \omega^2 x^2}} \,.$$

The integration yields:

$$\int dt = t - t_0 = \int \frac{dx}{\sqrt{c - \omega^2 x^2}} \,,$$

A second integration constant, "$-t_0$" is introduced; it too is determined by the initial conditions.

The integral over x looks tough. However, by a substitution of $x = (\sqrt{c}/\omega)\sin z$, it becomes easier to solve. After the substitution, the differential is, $dx = (\sqrt{c}/\omega)dz \cos z$. The integrand also simplifies: $\sqrt{c - \omega^2 x^2} = \sqrt{c - c\sin^2 z} = \sqrt{c}\sqrt{\cos^2 z} = \sqrt{c}\cos z$. The integral then becomes:

$$t - t_0 = \frac{\sqrt{c}}{\omega}\frac{1}{\sqrt{c}} \int \frac{dz \cos z}{\cos z} = \frac{1}{\omega} \int dz = \frac{z}{\omega} \,.$$

Multiplying the result by ω and taking the "*sine*" of both sides of the equation

60

yields:

$$\sin \omega(t - t_0) = \sin z = \frac{\omega}{\sqrt{c}} x \ .$$

Hence,

$$x(t) = A \sin(\omega t + \epsilon) \ ,$$

and two new constants, $A = \sqrt{c}/\omega$, and $\epsilon = -t_0\omega$, were defined simply because this is a customary way to display the answer to this problem. The appearance of the two integration constants in the solution is to be expected because the force only specifies the acceleration. One of the two constants accounts for the initial velocity and the other for the initial position. Using the specified initial conditions, $x(0) = 0$ and $v(0) = v_0$ yields,

$$0 = A \sin \epsilon \ , \qquad \text{and} \qquad v_0 = \frac{dx}{dt}\Big|_{t=0} = A\omega \cos \epsilon \ .$$

Hence, $\epsilon = 0$ and $A = v_0/\omega$. Therefore,

$$x(t) = \frac{v_0}{\omega} \cos \omega t \ .$$

EXERCISE 8.1

A bicyclist is riding along a straight flat road. If the angular displacement of the tires varies as: $\theta(t) = (3.00\,\text{rad/s})t + (0.650\,\text{rad/s}^3)t^3$, calculate the angular velocity of the tires.

EXERCISE 8.2

A clothes drier is running with a constant angular velocity of $400\,\text{rad/s}$. A sock gets stuck in the barrel of the drier and spins with the drier at $400\,\text{rad/s}$. Calculate the angular displacement of the sock as a function of time.

EXERCISE 8.3

A kitten tries to chase its tail and starts running in a circle. If its angular velocity is: $\omega(t) = 0.682\,\text{rad/s} + (1.26\,\text{rad/s}^2)t$. Calculate the number of rotations the kitten makes in the first four seconds.

EXERCISE 8.4

A compact disk player is not working properly. It doesn't spin the CD at a constant angular velocity. If the angular acceleration of the CD is: $\alpha(t) = (8.00\,\text{rad/s}^2)\sin((12.0\,\text{s}^{-1})t)$, calculate the angular velocity of the CD as a function of time. Assume that the initial angular velocity of the CD is $38.0\,\text{rad/s}$.

EXERCISE 8.5

A child decides to see how long she can spin in circles. She starts from rest, and her angular velocity is described by: $\omega(t) = 2.50\,\text{rad/s} + (7.50\,\text{rad/s}^2)t$. If she makes eight complete rotations before she falls over, what is her angular velocity just before she falls down?

EXERCISE 8.6

A car has a tire diameter of 16.0 inches. Its tires have an angular acceleration of: $\alpha(t) = -(1.20\,\text{rad/s}^3)t$. If its initial angular velocity is $30.0\,\text{rad/s}$, calculate the time it takes for the car to stop and the total distance the car travels before stopping.

EXERCISE 8.7

A child of mass $30.0\,\text{kg}$ is on a merry-go-round that has a mass of $100\,\text{kg}$, a radius of $1.30\,\text{m}$ and thickness of $0.100\,\text{m}$. Calculate the moment of inertia of the combination as a function of time, if the child walks at a constant rate directly toward the inside of the merry-go-round. Assume the radius of the position of the child is: $r(t) = 1.30\,\text{m} - (0.200\,\text{m/s})t$. Calculate the moment of inertia of the system (merry-go-round and child) as a function of time. Does the merry-go-round spin faster or slower as the child goes toward the center?

EXERCISE 8.8

Three squirrels are running around on a merry-go-round that was left spinning by some children. The angular velocity of the "system" of squirrels and the merry-go-round changes with time as a consequence of the moment of inertia of the "system" changing as: $I(t) = \dfrac{1{,}000 \text{ kg m/s}^2}{20 \text{ rad/s}-(2.0 \text{ rad/s}^2)t}$. Calculate the angular acceleration of the "system."

EXERCISE 8.9

A hamster, running on an exercise wheel, exerts a torque on the wheel. If the wheel has an angular velocity that can be expressed as:

$$w(t) = 3.0 \text{ rad/s} + (8.0 \text{ rad/s}^2)t + (1.5 \text{ rad/s}^4)t^3,$$

calculate the torque on the wheel as a function of time. Assume that the moment of inertia is $500 \text{ kg} \cdot \text{m}^2$ and is constant.

EXERCISE 8.10

A person is trying to spin a frisbee on his finger, and does not keep his finger directly centered on the frisbee. His finger creates a torque on the frisbee, due to friction. If the position of his finger relative to the center of rotation varies as: $r(t) = (1.00 \text{ cm}) \sin((3.00 \text{ s}^{-1})t)$, calculate the angular momentum of the frisbee as a function of time. Assume the coefficient of friction is 0.30, the mass of the frisbee is 175 g, and that the frictional force is independent of the speed of rotation. The initial angular momentum of the frisbee is $0.020 \text{ kg} \cdot \text{m}^2/\text{s}$.

Chapter 9 ENERGY

EXAMPLE 9.1

The following simple model can be used to explain the origin of the friction force acting on a macroscopic object moving through a gas at high speed. Imagine an object, lets say cylindrical in shape, with the base of the cylinder having the area A. The object is moving through a gas at high speed v in a direction of the normal to the base of the cylinder. The molecules (or atoms) of the gas through which the object is moving can be for practical purposes considered as stationary. Assume that the moving object collides elastically with molecules of the gas it encounters. By calculating the rate at which the energy is transferred to the molecules as a result of collisions find the expression for the friction force. Assume that the density of the gas is ρ.

SOLUTION

[Given: a macroscopic object of cylindrical shape with base of area A moving with speed v in the direction perpendicular to its base. The gas molecules prior to collision are stationary and collisions are elastic. The density of the gas is ρ. Find: the friction force.]

There are two parts to this problem. In the first part we have to figure out how does the energy transfer works. Then we can calculate the amount of energy transferred per unit time. From there we will be able to obtain the friction force.

To learn how does the mechanism of the energy transfer works, let us consider a perfectly elastic collision between a macroscopic object moving at high speed and a tiny, microscopic object – molecule, that is stationary. For simplicity, let us consider head-on collisions. Because the collision is elastic (the assumption) the energy is conserved during the collision. Let M be the mass of the cylinder and let m be the mass of a single molecule. The speed of the cylinder before the collision is v and the speed of the cylinder after the collision is v'. The speed of the molecule before the collision is 0 and after the collision is v_m. To find v' and v_m, use the laws of the conservation of momentum and energy. Because there are no external or internal forces acting on the system, the energy is just the kinetic energy. The momentum and energy conservation laws read:

$$Mv = Mv' + mv_m , \qquad \text{and} \qquad \frac{Mv^2}{2} = \frac{Mv'^2}{2} + \frac{mv_m^2}{2} .$$

In solving this pair of equations it is best to first combine the equations so that we do not have to deal with the quadratic formula. Rewrite the conservation laws as

$$M(v - v') = mv_m , \qquad M(v^2 - v'^2) = M(v - v')(v + v') = mv_m^2 .$$

In rewriting the energy conservation law, the formula $a^2 - b^2 = (a - b)(a + b)$ was utilized. The factor $M(v - v')$ in the energy equation can be replaced by mv_m on the account of momentum conservation law, yielding,

$$v + v' = v_m .$$

This equation and the momentum conservation law constitute a pair of linear equations that are solved for velocities v' and v_m. Substituting the $v_m = v + v'$ into the momentum conservation equation yields

$$M(v - v') = m(v + v') \qquad \text{and} \qquad v' = \frac{M - m}{M + m} v .$$

Using this result gives,

$$v_m = v + \frac{M - m}{M + m} v = \frac{2M}{M + m} v .$$

Next, use the fact that the cylinder is macroscopic and that the molecule is microscopic, $M \gg m$. Hence, $\frac{M-m}{M+m} \approx 1$ and $\frac{M}{M+m} \approx 1$. This implies that $v' \approx v$ and

$v_m \approx 2v$. In other words, the velocity of the cylinder is practically unchanged after the collision and the molecule of the gas rushes away from the cylinder with speed $2v$ in the direction of the motion of the cylinder. The energy of the molecule after the collision is

$$\text{KE}_m = \frac{mv_m^2}{2} \approx \frac{m(2v)^2}{2} = 2mv^2 \ .$$

The energy of the molecule after the collisions also equals to the energy lost by the cylinder as a result of a single collision.

Let us now calculate the rate at which the cylinder looses energy. Observe the motion of the cylinder during a short time interval Δt. From the analysis of the collision process we know that the speed of the cylinder practically remains unchanged as a result of a single collision. Hence, the time Δt should be sufficiently short so that not too many collisions occur. We can assume that Δt is infinitesimal. During the time interval Δt the cylinder moves a distance $v\Delta t$. Therefore, all of the molecules on the path of the cylinder contained inside the volume $\Delta V = Av\Delta t$ will collide with the cylinder. Since the density of the gas is ρ, the total mass of all the molecules that collide with the cylinder during the time interval Δt is $\Delta m = \rho \Delta V = Av\rho \Delta t$. Since the mass of the single molecule is m, the total number of molecules inside the volume ΔV is $\Delta m/m = Av\rho \Delta t/m$. Using the fact that the energy taken away by a single molecule during the collision is $2mv^2$, the energy lost by the cylinder due to all collisions it experienced during the time interval Δt is:

$$\Delta \text{E} = -2mv^2 \frac{Av\rho \Delta t}{m} = -2A\rho v^3 \Delta t \ .$$

Hence, the cylinder is loosing the energy (negative sign) at a rate

$$\frac{\Delta \text{E}}{\Delta t} = -2A\rho v^3 \ .$$

Therefore, $\Delta \text{E}/\Delta t$ is the power lost by the cylinder as it is moving through the gas. It is equal to the work per unit time done by the friction force F_f acting on the cylinder. Hence

$$\frac{\Delta E}{\Delta t} = P_f = F_f v \ .$$

By comparison, the friction force is:

$$F_f = -2\rho A v^2 \ .$$

The friction force is proportional to the square of the speed of the moving object. The force coefficient, $2\rho A$, depends on the density of the gas and on the cross-section of the object. This result makes sense: if the gas is denser there will be more molecules to collide with, hence more energy loss; if the cross-section is bigger, there will be more collisions, hence more energy loss. Also, if the speed is greater there will be more energy taken away per collision, hence, more total energy loss.

EXAMPLE 9.2

A chain of total length L and mass M lies on the table-top. At time $t = 0$ chain is picked up and lifted from the table straight up, gradually, at constant speed v. Find the work done by the lifting force until all of the chain has been lifted from the table.

SOLUTION

[Given: a chain of length L and mass M. At $t = 0$ the chain is picked up and gradually lifted straight up with speed $v = $ constant. Find: the work W done by the lifting force until all of the chain is lifted from the table.]

First find the force that does the work. Let $\lambda = M/L$ be the linear mass density of the chain. The lifting force performs two actions – it supports the weight of the part of the chain that has already been lifted from the table and it is also responsible for bringing into motion at constant speed the part of the chain that was originally motionless on the table top. Let the length of the chain already in the air at time $t > 0$ be s. Then the weight of this part of the chain is $\lambda s g = M g s / L$. Let also, Δs be the length of small segment of the chain that has been elevated from the table top during the time interval Δt. Before it was elevated from the table top, the chain segment of of length Δs had no momentum. Once it has been elevated into the air, the chain segment is moving with constant speed v. Hence its momentum is, $\lambda v \Delta s = M v \Delta s / L$. The momentum change per unit time is $(M v \Delta s / L) / \Delta t = M v^2 / L$. In writing the last equation we used the fact $\Delta s / \Delta t = v$. Therefore, the total lifting force at time t is $F = \frac{M}{L}(g s + v^2)$. Since the chain is traveling at constant speed, $s = v t$. The force F acts on the chain until all of the chain is lifted from the table top. Hence, the total time during which the force F acts is $T = L / v$.

Next, let us calculate the work done by the force F. By definition, the work done by the force equals to the integral of the force over a path along which the object on which the force acts is moving. In our case the motion begins on the table top and the force stops acting when all of the chain is in the air. Let s_{max} be the distance that the chain covered until the force F ceased to act. Note that it is not *a priori* clear what to take for s_{max}. However it is easy to see how long will the force F act on the chain. Hence, we write

$$W = \int_0^{s_{max}} F \, ds = \int_0^T F \frac{ds}{dt} dt \; .$$

In writing the second equality we have converted the conventional integral that is used to define the work into a more convenient expression (it this problem) involving the integration over time because the force is time dependent.

We must proceed carefully becasue the force has two parts which perform different action. Let us write $W = W_1 + W_2$. Let W_1 be the work of the part of the force that supports the weight of the part of the chain already in the air. Let W_2 be the work of the part of the force that changes a momentum of a tiny portion of the chain that is coming of the table.

It is easy to calculate W_1.

$$W_1 = \int\limits_0^{L/v} \frac{M}{L} g v^2 t\, dt = \int\limits_0^L \frac{M}{L} g s\, ds = Mg\frac{L}{2}.$$

Clearly, the work W_1 represents the change in the potential energy of the center of mass of the chain. When the chain is on the table top it has no potential energy (using the surface of the table as the reference point for measuring the gravitational potential energy). When all of the chain is in the air, its center of mass is at a height $L/2$ and the potential energy is $mgL/2$.

To calculate the work W_2 is tricky. The constant term in the force, Mv^2/L, acts only at the point of contact between the chain and the table. Hence,

$$W_2 = \int\limits_0^{L/v} \frac{M}{L} v^2 \frac{ds}{dt}\, dt \ .$$

Setting $ds/dt = v$ which may be your gut choice, leads to a wrong answer, $W_2 = mv^2$.* The correct choice is $ds/dt = v/2$. Why? First, note that the integrand in the formula for the work W_2 is the power expended by the force. Next, observe an infinitesimal segment of the chain during an infinitesimal amount of time as it is coming of the table top and lets calculate the power expended by the force that changes the momentum of that segment of the chain. During that infinitesimal

---

* To see that Mv^2 is an incorrect answer consider observing the chain just when it is of the table. It is moving with constant speed and its total energy is $MgL/2 + Mv^2/2$. However, just as the last piece is coming of the table top, if Mv^2 is correct, its total energy would be $Mv^2 + MgL/2$. Clearly, an amount of energy $Mv^2/2$ suddenly disappears into a thin air. This can not happen because the energy is conserved. So, Mv^2 is incorrect.

amount of time the segment of the chain changes its state from being stationary on the table top to moving above the table at speed v. Hence, this infinitesimal chain segment must be accelerated from zero speed to a final speed v during an infinitesimal time interval. Its average speed (assuming constant acceleration during the infinitesimal time interval to make the argument easier) is $v/2$. Hence, the power expended by the force is $(Mv^2/L)(ds/dt) = (Mv^2/L)(v/2)$, or $ds/dt = v/2$. Therefore,

$$W_2 = \frac{Mv^2}{L} \int_0^{L/v} \frac{v}{2} dt = \frac{1}{2}Mv^2 .$$

The total work done by the lifting force F is

$$W = \frac{1}{2}Mv^2 + Mg\frac{L}{2} .$$

EXAMPLE 9.3

Consider building a uniform, massive, spherical object having total mass M and radius R by bringing in small pieces of the mass from the infinity. What is the total energy (gravitational) required?

SOLUTION

[Given: total mass M of the uniform object being built and its final radius R. Find: total (gravitational) energy required if building blocks are brought from the infinity.]

Let us imagine that the spherical object being built has been partially completed and it has the radius $r < R$. At that moment its mass is $m = \rho V_r$, where $\rho = \frac{M}{\frac{4\pi}{3}R^3}$ is the mass density of the object, and $V_r = \frac{4\pi}{3}r^3$ is the volume of the completed portion. Imagine now bringing in an additional small piece of mass from infinity.

The mass of the added part is $dm = \rho dV$. The required work is equal to the potential energy that the small massive piece has once it is put in place:

$$dW = Gm\frac{dm}{r} = \frac{3GM^2r^2}{R^6}r^2dr \ ,$$

where we used the spherical symmetry of the problem and wrote $dV = 4\pi r^2 dr$, a form appropriate for the spherical shell of radius r and thickness dr.* Integrating the work required to bring one small piece over the radial distance from $r = 0$ to $r = R$ we get the total work:

$$W = \int\limits_{0}^{R} \frac{3GM^2}{R^6}r^4dr = \frac{3GM^2}{R^6}\frac{R^5}{5} \ .$$

Hence,

$$W = \frac{3}{5}\frac{GM^2}{R} \ .$$

This result is quite interesting. It can be interpreted as the total gravitational energy stored in a massive body of mass M and radius R.

EXERCISES

EXERCISE 9.1

At a construction site a worker pushes horizontally against a large bucket filled with some construction material. The total mass of the bucket is $800\,\mathrm{kg}$. The bucket is suspended from crane by a $30\,\mathrm{m}$ long cable. How much work has to be done on the bucket to move it slowly $1.5\,\mathrm{m}$ away from the vertical? How much work is done by the gravity?

---

* You can also start with $dV = r^2 dr \sin\theta d\theta d\phi$, where θ and ϕ are the polar angles in the three-dimensional space, $0 \leq \theta \leq \pi$, $0 \leq \phi \leq 2\pi$. Integrating over the angles yields $\int_0^\pi \sin\theta d\theta \int_0^{2\pi} d\phi = 4\pi$.

EXERCISE 9.2

A particle moves along the x-axis under the influence of variable a force $F(x)$. Show that the force $F(x)$ is conservative.

EXERCISE 9.3

A particle moves along the x-axis under the influence of a variable force $F(x) = (5\,\text{N/m}^2)x^2 + (3\,\text{N/m})x$. What is the potential energy associated with the force F? Assume that the potential energy takes the value 0 at the origin.

EXERCISE 9.4

A particle moves in a plane under the influence of conservative force $\mathbf{F} = ay\hat{x} + ax\hat{y}$. What is the work done by this force as the particle moves from the origin (point with $x = 0$ and $y = 0$) to the point with coordinates $x = x_0$ and $y = y_0$? What is the potential energy associated with the force $\mathbf{F}$?

EXERCISE 9.5

The potential energy of the atom in a Hydrogen molecule H_2 may be taken to be $U(r) = U_0(e^{-2(r-r_0)/b} - 2e^{-(r-r_0)/b})$ where $U_0 = 4.484 \times 10^{-19}$ J and $r_0 = 0.37 \times 10^{-10}$ m and $b = 0.34 \times 10^{-10}$ m. The distance r is measured from the center of the molecule. Within the molecule, the atom moves and experiences the force due to the molecular potential energy. Find the force acting on the atom. Find the location within the molecule where the force is zero.

EXERCISE 9.6

Consider a projectile traveling horizontally and is slowing down under the influence of air resistance. The mass of the projectile is 50 kg and the speed is a function of time given by $v(t) = b - ct + dt^2$, where $b = 700$ m/s, $c = 62$ m/s^2 and $d = 3$ m/s^3. (a) What is the instantaneous power removed from the projectile by the air resistance? (b) What is the kinetic energy at $t = 0$ s and at $t = 3.0$ s. (c) What is the average

72

power consumed by the air resistance during the time interval from $t = 0\,$s to $t = 3.0\,$s.

EXERCISE 9.7

Calculate the potential energy of a mass m inside the uniform spherical shell of radius R and total mass M.

EXERCISE 9.8

A force applied on the object varies with position according to the formula $F = (2\,\text{N/m}^3)x^3 - 4$. How much work is done by this force on an object that moves from $x = 2\,$m to $x = 5\,$m.

EXERCISE 9.9

A block of mass $3\,$kg starts from rest $60\,$cm above the ground on an incline plane with angle of 30^0. The coefficient of the friction between the block and the incline surface $\mu_k = 0.20$. Once the block has reached the bottom of the incline it continues along the horizontal until it stops because of the friction. How far will the block travel along the horizontal surface if the coefficient of friction for the block moving along the horizontal surface is the same as for the incline?

Chapter 10   SOLIDS

EXAMPLE 10.1

A child jumps on a trampoline and springs up. Treat the trampoline as a spring. When the child stands stationary on the trampoline he depresses it 3.00 cm due to his weight (his mass is 40.0 kg). When he jumps onto it, the maximum depression of the trampoline is 12.0 cm. Calculate the boy's initial velocity (just before he hits the trampoline), and calculate the boy's maximum height. Assume there is no energy lost.

SOLUTION

[Given: $y_0 = -3.00$ cm, $y_{min} = -12.0$ cm, $m_b = 40.0$ kg. Find: v_i, y_{max}.]

First, we know the force due to the spring is $ky - up$, where y is the distance of compression of the spring. With the boy standing on the trampoline we can determine the spring constant. The force exerted on the spring is the boy's weight, so $F = mg = ky_0$ or

$$k = \frac{mg}{y_0} = \frac{(40.0\,\text{kg})(9.81\,\text{m/s}^2)}{0.0300\,\text{m}} = 1.31 \times 10^4\,\text{kg/s}^2.$$

We can then determine the potential energy stored in the spring when it is fully compressed because it equals the work done by the boy jumping down. The work of the spring force is given by the integral of the force over the distance compressed,

$$W = \int F\,dy = \int_0^{-12.0\text{cm}} ky\,dy = \frac{1}{2}ky^2 \Big|_0^{-12.0\text{cm}}.$$

Hence, the potential energy which is stored in the spring is $\text{PE}_s = \frac{1}{2}ky^2\Big|_{0\text{cm}}^{-12.0\text{cm}} = \frac{1}{2}(1.31 \times 10^4\,\text{kg/s}^2\left[(-0.12\,\text{m})^2 - (-0.00\,\text{m})^2\right] = 94\,\text{J}$. Notice that the force of the boy is down, while the force of the spring is up.

Now we need to use the conservation of energy to determine the boy's initial velocity and his maximum height. $E_{total} = KE + PE_e + PE_G$, where PE_e is the elastic potential energy, PE_G is the gravitational potential energy, and E_{total} is the total energy which is conserved. Explicitly, we have $KE = \frac{1}{2}mv^2$, $PE_e = \frac{1}{2}ky^2$, and $PE_G = mgy$. We have three different conditions to examine: (a) Just as the boy hits the trampoline, $(y = 0, v = v_i)$, (b) when the boy is 12.0 cm below the starting position (maximum depression) of the trampoline, $(y_{min} = -12.0\,\text{cm}, v = 0)$, and (c) the maximum height of the boy, $(y = y_{max}, v = 0, PE_s = 0)$. The total energy in all situations is the same (conservation of energy): $E_a = E_b = E_c$. We have

$$E_a = \tfrac{1}{2}mv_i^2; \qquad E_b = \tfrac{1}{2}k(-12.0\,\text{cm})^2 + mg(-12.0\,\text{cm}); \qquad E_c = mgy_{max}$$
$$= 47.1\,\text{J}.$$

Equation $E_a = E_b$ enables us to find the initial velocity v_i. Equation $E_b = E_c$ enables us to find the maximum height y_{max}. We have:

$$\tfrac{1}{2}mv_i^2 = E_b = 47.1\,\text{J} \qquad \text{or} \qquad v_i = \sqrt{\tfrac{2E_b}{m}} = \sqrt{\tfrac{2(47.1\text{J})}{40.0\text{kg}}} = 1.53\,\text{m/s}\ .$$

$$mgy_{max} = E_b = 47.1\,\text{J} \text{ or } y_{max} = \frac{E_b}{mg} = \frac{47.1\,\text{J}}{(40.0\,\text{kg})(9.81\,\text{m/s}^2)} = 0.120\,\text{m} = 12.0\,\text{cm}\ .$$

In this case, the boy's final height is the same as the distance he depresses the trampoline. What if his initial velocity is increased to 2.00 m/s? The trampoline would have a maximum depression of 14.4 cm and his final height would be 20.0 cm. Try it!

EXAMPLE 10.2

Find the work necessary to compress a rubber eraser, confined in a cylindrical plastic case. Assume the stress goes as $\sigma = (1.00 \times 10^6 \, \text{Pa})\epsilon^{1/2}$, where ϵ is the strain. The original length of the eraser is $10.0 \, \text{cm}$, its cross-sectional area is $A = 0.800 \, \text{cm}^2$, and the eraser is compressed a distance of $0.500 \, \text{cm}$.

SOLUTION

[Given: $\sigma = C\epsilon^{1/2}$, where $C = 1.00 \times 10^6 \, \text{Pa}$; $L = 10.0 \, \text{cm}$; $A = 0.800 \, \text{cm}^2$; and $\Delta L = 0.500 \, \text{cm}$. Find: W.]

Remember the work done is $W = \int F \cdot dl$ (from Eq. (9.6) in the textbook), so we know we need to integrate the force over the 0.5 cm distance of compression. If the force was constant with distance compressed, then $W = F \cdot l$, but that is not the case in this problem. We now need to get an expression for the force. Since $\sigma = C\epsilon^{1/2}$, and $\sigma = \frac{F}{A}$ and $\epsilon = \frac{\Delta L}{L}$, we know $\frac{F}{A} = C(\frac{\Delta L}{L})^{1/2}$ or $F = CA(\frac{\Delta L}{L})^{1/2}$. In this equation, A is the cross-sectional area, which is constant because the eraser is confined in a cylindrical case. L is the original length of the eraser, and ΔL is the distance the eraser is compressed. So, to calculate the work done, dl is actually $d(\Delta L)$, so that

$$W = \int_0^{0.5\,\text{cm}} F \cdot d(\Delta L) = \int_0^{0.5\,\text{cm}} CA \left(\frac{\Delta L}{L}\right)^{1/2} d(\Delta L) = \frac{2}{3}CA\frac{1}{L^{1/2}}(\Delta L)^{3/2}\Bigg|_{\Delta L = 0\,\text{cm}}^{\Delta L = 0.5\,\text{cm}}$$

$$= \frac{2}{3}(1.0 \times 10^6 \, \text{Pa})\left(\frac{0.8\,\text{cm}^2}{\sqrt{10\,\text{cm}}}\right)\left[(0.5\,\text{cm})^{3/2} - (0\,\text{cm})^{3/2}\right] = 5.96\frac{\text{N}}{\text{m}^2} \cdot \text{cm}^3 \left(\frac{1\,\text{m}}{100\,\text{cm}}\right)^3$$

$$W = 5.96 \times 10^{-2} \, \text{N·m} = 5.96 \times 10^{-2} \, \text{J} .$$

So the work required to compress the rubber eraser by $0.5 \, \text{cm}$ is $5.96 \times 10^{-2} \, \text{J}$. Notice that the farther you try to compress the rubber eraser the more work that is necessary, which should seem reasonable, but the work increases faster than the distance compressed. That should also seem reasonable. The eraser is easy to compress a little, but the more you try to compress it the harder it gets.

EXAMPLE 10.3

Consider the slab shown in figure **EG-10.3a**. It has a height of 4.00 cm, a minimum width of 0.200 cm, a maximum width of 2.20 cm and a depth of 6.50 cm. The base of the slab is glued to a table and cannot move. A force of 21.7 N is applied horizontally at the top (narrowest) part of the slab.

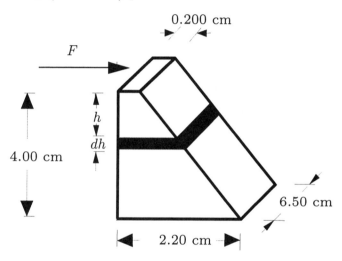

Figure: EG-10.3a

The slab is made of rubber and has a shear modulus of 0.800×10^6 Pa which does not change with applied force or the distance displaced. Calculate the amount of shift that occurs at the top (narrowest) part of the slab.

SOLUTION

[Given: $H = 4.00$ cm, $w_{min} = 0.200$ cm, $w_{max} = 2.20$ cm, $D = 6.50$ cm, $F = 21.7$ N, $S = 0.800 \times 10^6$ Pa. Find: ΔL.]

The shear modulus is defined as $S = -\frac{\sigma_S}{\epsilon_S}$, where $\epsilon_S = \frac{\Delta L}{L_0}$ is the strain and $\sigma_S = \frac{F}{A}$

is the stress. We can use these relations to find the compression: $\Delta L = \frac{FL_0}{SA}$. Now consider a small section of the slab, as shown in the figure. It has a height dh and an area $A = D[w_{min} + \frac{h}{H}(w_{max} - w_{min})]$. We next need to consider the force applied to this small section of the slab. Each section of the slab is in equilibrium, and therefore does not experience a net force. There are, however, forces acting on each section of the slab. First consider the upper most section. The external force, F, acts to the right, so there must be a force of equal magnitude acting to the left, see figure **EG-10.3b**. This force is generated by the next lower section, and is internal to the slab. Therefore, Newton's third law says, there must be another force acting to the right on the second section. Pairs of these internal forces act between each slice of the slab, until we reach the bottom. The bottom slice feels a force to the right from above, and the glue creates a force to the left from below. So every slice feels a force of magnitude $21.7\,\text{N}$ on the top surface to the right and an equal force of $21.7\,\text{N}$ on the bottom surface to the left, thereby keeping the entire slab in equilibrium.

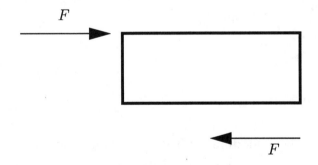

Figure: EG-10.3b

Also notice that our equation contains a variable, L_0. L_0 is the unstretched length (height) of the small slab, or $L_0 = dh$. So our equation now becomes:

$$d(\Delta L) = \frac{-F \cdot dh}{S \cdot D[w_{min} + \frac{h}{H}(w_{max} - w_{min})]} \; .$$

Now, if we integrate over all the small slabs to get the expression for ΔL, which is the final shift of the top of the slab:

$$\int_{\Delta L(h)}^{0} d(\Delta L) = \frac{-F}{S \cdot D} \int_{0}^{H} \frac{dh}{\left[w_{min} + \frac{h}{H}(w_{max} - w_{min})\right]} \qquad \text{(integrating top down)}$$

To do this integration we make a substitution of $x = w_{min} + \frac{h}{H}(w_{max} - w_{min})$. So that $dx = \frac{(w_{max} - w_{min})}{H} \cdot dh$, or $dh = \frac{H \, dx}{(w_{max} - w_{min})}$. Substituting gives us:

$$-\Delta L = \frac{-F}{S \cdot D} \cdot \frac{H}{(w_{max} - w_{min})} \int_{w_{min}}^{x = (w_{max} - w_{min}) + w_{min}} \frac{1}{x} \, dx$$

$$\Delta L = \frac{F \cdot H}{S \cdot D(w_{max} - w_{min})} \int_{w_{min}}^{w_{max}} \frac{1}{x} \, dx$$

$$= \frac{F \cdot H}{S \cdot D(w_{max} - w_{min})} \ln(x) \Big|_{w_{min}}^{w_{max}}$$

$$= \frac{F \cdot H}{S \cdot D(w_{max} - w_{min})} \ln\left(\frac{w_{max}}{w_{min}}\right)$$

$$\text{recall that } \ln(A) - \ln(B) = \ln\left(\frac{A}{B}\right)$$

$$\Delta L = \frac{(21.7\,\text{N})(4 \times 10^{-2}\,\text{N})}{(0.8 \times 10^{6}\,\text{Pa})(0.065\,\text{m})(0.02\,\text{m})} \ln\left(\frac{0.022\,\text{m}}{0.002\,\text{m}}\right)$$

$$= 2.00 \times 10^{-3}\,\text{m} = 0.200\,\text{cm}$$

This tells us that the top of the rubber slab moved a distance of 0.200 cm relative to its position before it was acted on by the force. If this problem was done with a square slab of constant width, $w = 2.20$ cm, the top of the slab would only move 0.076 cm. Try it! This slab bends more because its thinner at the top.

EXERCISE 10.1

A slinky hangs from the ceiling and has a mass per unit length given by $\mu(l) = (0.25\,\text{kg}\,\text{m})(l + 0.50\,\text{m})^{-2}$, where $l = 0.0\,\text{m}$ at the lowest point on the slinky and $l = 1.0\,\text{m}$ at the ceiling. Calculate the total mass of the slinky.

EXERCISE 10.2

Frost is forming in an orange grove. A spherical orange bud has frost building up on its entire spherical surface at a rate of 0.10 mm/hr. Given that the average volume of a water molecule is $\frac{4}{3}\pi(3.0\times10^{-10}\,\text{m})^3$, find the rate $\frac{dN}{dt}$, at which water molecules are accumulating when $R = 12\,\text{mm}$.

EXERCISE 10.3

The mass per unit height of an ice cream cone varies as $\mu = \mu_0 2\pi r$, where r is the radius of the ice cream cone and the total height is $12\,\text{cm}$. The angle of the ice cream cone α, as shown in the figure is $14°$. If the total mass of the ice cream cone is $15\,\text{g}$, what is the value of μ_0.

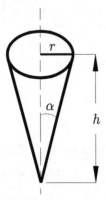

Figure: EX-10.3

EXERCISE 10.4

Calculate the change in the density as a function of time of whipped cream in a spray can. Originally, 100 g of whipped cream was in the 1-liter can. When it is sprayed, the whipped cream leaves the can at a rate of 0.50 g/s. How long does it take for a person to reduce the density in the can to half of its original value?

EXERCISE 10.5

The bulk modulus for steel is $B = 160 \times 10^9$ Pa. If a $1.00 \, \text{m}^3$ block of steel is submerged in the ocean, where the pressure due to water depends on depth as: $P(h) = 1.01 \times 10^5 \, \text{Pa} + (1.005 \times 10^4 \, \text{Pa/m})(h)$, calculate the change in the fractional volume of the steel as a function of the depth in the seawater, $\frac{d}{dh}\left(\frac{\Delta V}{V_0}\right)$.

EXERCISE 10.6

Consider a solid copper sphere submerged in a large beaker of benzene. Calculate the change in the radius of the sphere as a function of depth, if the sphere has mass of 40.0 g and an initial volume of $4.50 \, \text{cm}^3$. The pressure due to benzene depends on depth as: $P(h) = 1.01 \times 10^5 \, \text{Pa} + (8.61 \times 10^3 \, \text{Pa/m})h$, and the bulk modulus of copper is 140×10^9 Pa.

EXERCISE 10.7

The bulk modulus for aluminum is $B_{Al} = 70.0 \times 10^9$ Pa. If a 100 g sphere of aluminum is submerged in distilled water, where the pressure due to the water depends on the depth as: $P(h) = 1.01 \times 10^5 \, \text{Pa} + (9.80 \times 10^3 \, \text{Pa/m})h$, calculate the change in the density of the aluminum as a function of the depth in distilled water, $\frac{d}{dn}(\rho)$. The initial density of the aluminum was $2.70 \, \text{g/cm}^3$.

EXERCISE 10.8

A child's toy consists of a wooden paddle which has a rubber ball attached to it by an elastic string. After the paddle hits the ball, the ball stretches the elastic an additional 6.0 cm before it starts to return. The stress-strain profile is given by

$$\sigma = (1.0 \times 10^6 \, \text{Pa})\epsilon^{1/2} + \frac{1}{16}(1.0 \times 10^6 \, \text{Pa})\epsilon^2.$$

Find the work done by the ball to stretch the elastic. Assume the cross sectional area of the elastic string is $0.10 \, \text{cm}^2$ and the length of the unstretched elastic is 35 cm.

EXERCISE 10.9

At lunchtime in the elementary school cafeteria, a bowl of mashed potatoes is placed on a boy's chair the instant before he sits down. When he sits down he smushes the mashed potatoes a distance of 0.600 cm. The boy is a trendsetter, and soon it is happening to everyone. They find that the work required to smush the mashed potatoes depends on how far they are smushed: $W(h) = (2.15 \times 10^5 \, \text{J}) \left(\frac{h}{H}\right)^{3/2}$, where h is the distance smushed and H is the original height of the potatoes. Assume that the bowl has a cross-sectional area of $700 \, \text{cm}^2$ and a height of 8.00 cm (the bowl is a cylinder). Calculate the form of the stress-strain profile for the mashed potatoes.

EXERCISE 10.10

A roll of pizza dough hangs under its own weight. When rolled on the table (before it is hung up) it has a cross-sectional area of $A = 0.20 \, \text{m}^2$, a mass per unit length $\mu = 0.28 \, \text{kg/m}$ (which remains constant) and a length of $L = 1.6 \, \text{m}$. Assume the cross-sectional area remains the same and that Young's modulus (which remains constant) for pizza dough is: $Y = 8.0 \times 10^4 \, \text{Pa}$, find the length of the hanging pizza dough.

EXAMPLE   11.1

A water in a big reservoir lake is held by a dam of width of 100 m. If the water in the reservoir just behind the dam is 30 m deep what is the total force acting on the dam?

SOLUTION

[Given: the dam of width $w = 100$ m; the water just behind the dam is $H = 30$ m deep. Find: total force on the dam.]

The geometry of the problem is given in Fig. EG-11.1. Also, useful variables are identified.

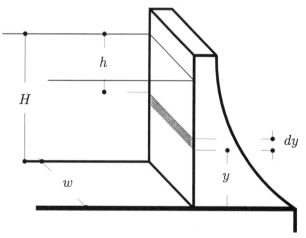

Figure: EG-11.1

Assuming static situation, the force on the dam is a result of the hydrostatic pressure of the water in the reservoir. As a function of the depth h the pressure is given by $P = \rho g h$ where $\rho = 1,000$ kg/m^3 is the density of water. To find the total force on the dam we need to calculate the integral of the pressure over the portion of the surface of the dam that is under the water. Note that the atmospheric pressure can be ignored because it affects the water and the dam equally. Dividing the area of the dam into rectangular segments of height dy and width w (see Fig. EG-11.1), the differential element of the area is $dA = wdy$. Hence, the total force is given by

$$F = \int_0^H P dA = \int_0^H \rho g h w dy .$$

The depth h can be expressed in terms of the water level H and the distance y measured from the bottom of the dam, $h = H - y$. Hence,

$$F = wg\rho \int_0^H (H - y) dy = wg\rho \left(H^2 - \tfrac{1}{2}H^2\right) = \tfrac{1}{2}wg\rho H^2 = \tfrac{1}{2}g\rho AH ,$$

where the answer is expressed in terms of the flooded area $A = wh$ of the dam. Using the numerical values of the constants in the problem, we have

$$F = \frac{1}{2}(9.8\,\mathrm{m/s^2})(1000\,\mathrm{kg/m^3})(100\,\mathrm{m})(30\,\mathrm{m^2}) = 4.4 \times 10^8\,\mathrm{N} .$$

It is apparent from the answer that the force on the dam is large. For example, the weight of an average person of mass of about $60\,\mathrm{kg}$ is approximately $600\,\mathrm{N}$. The calculated force equals to the the total weight of about $730,000$ average persons! Notice that the force depends on the area that is flooded as well as on the depth. This means that an object of small area experiences a smaller force which grows with the depth.

EXAMPLE 11.2

Water flows through a constricted pipe depicted in Fig. EG-11.2. Assume that the water is incompressible and nonviscuous and that the flow itself is nonrotational and steady. At the point R_1 at the lower portion of the pipe the pressure is $P_1 = 2.5 \times 10^4$ Pa; at the point R_2 at the upper portion of the pipe the pressure is $P_2 = 1.5 \times 10^4$ Pa. The pipe diameter at the point R_1 is 8.0 cm and at the point R_2 the pipe radius is 4.0 cm. The point R_2 is elevated relative to the point R_1 by 50 cm. (a) Find the speed of the flow of the water in the upper and the lower end of the pipe. (b) Determine the rate at which the water flows through the pipe.

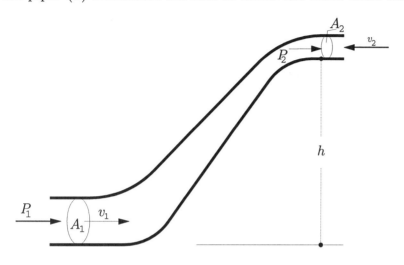

Figure: EG-11.2

SOLUTION

[Given: $P_1 = 2.5 \times 10^4$ Pa, $P_2 = 1.5 \times 10^4$ Pa; $r_1 = 8$ cm, $r_2 = 4.0$ cm; $h = 50$ cm. Find: (a) v_1 and v_2; (b) $\frac{dV}{dt}$.]

Let us observe the motion of the fluid through the pipe for a short time interval Δt. Let us also assume that the velocity of the fluid at the lower end of the pipe (point R_1) is v_1 and that the velocity of the fluid at the upper end (point R_2) is v_2. During the time Δt the fluid covers the distance $\Delta x_1 = v_1 \Delta t$ at R_1 and distance $\Delta x_2 = v_2 \Delta t$ at R_1. Since the cross-section of the pipe is circular, the area of the pipe at R_1 is $A_1 = r_1^2 \pi$ and at R_2 is $A_2 = r_2^2 \pi$. The volume of the fluid that passed through the area A_1 during the time interval Δt is $\Delta V_1 = A_1 \Delta x_1 = A_1 v_1 \Delta t$ and

85

the volume of fluid that passed through the area A_2 during the same time Δt is $\Delta V_2 = A_2 \Delta x_2 = A_2 v_2 \Delta t$. Because the flow is uniform the flow rate $Q = \Delta V / \Delta t$ is constant. Hence,

$$Q = \frac{\Delta V_1}{\Delta t} = \frac{\Delta V_2}{\Delta t} = \text{constant}$$

implying that $\Delta V_1 = \Delta V_2 = \Delta V$. We will need this result shortly.

The fluid entering the pipe segment at R_1 exerts the pressure P_1 to the right and the fluid that leaves the pipe segment at R_2 exerts the pressure P_2 to the left. Note that the directions of the pressures are dictated by Newton's third law and the fact that the fluid motion is steady, *i.e.* there is local equilibrium! This means that there is a force $F_1 = P_1 A_1$ acting to the right at point R_1 and there is a force $F_2 = P_2 A_2$ acting to the left at point R_2 . At the lower end (point R_1) the force F_1 does the work $W_1 = F_1 \Delta x_1$ and at the upper end (point R_2) the force F_2 does the work $W_2 = -F_2 \Delta x_2$. Hence, the net work of the forces acting on the fluid in the pipe is

$$\begin{aligned} W = W_1 + W_2 &= F_1 \Delta x_1 - F_2 \Delta x_2 \\ &= P_1 A_1 \Delta x_1 - P_2 A_2 \Delta x_2 = P_1 \Delta V_1 - P_2 \Delta V_2 \\ &= (P_1 - P_2) \Delta V \ . \end{aligned}$$

In writing the last equality we have used the fact that the flow rate is constant (see the paragraph above).

Since the pipe rises (the difference in heights between two ends is h) some of the work is consumed by changing the potential energy of the liquid. The remainder of the work is consumed by changing the kinetic energy of the fluid. Hence,

$$W = (P_1 - P_2) \Delta V = \Delta \text{KE} + \Delta \text{PE} \ .$$

The fluid is uniform and has density ρ, so,

$$\Delta \text{KE} = \frac{\Delta m v_2^2}{2} - \frac{\Delta m v_1^2}{2} \ .$$

Note that the fact that the the fluid flow is steady is used to write $\Delta m_1 = \rho \Delta V_1 =$

$\rho\Delta V = \rho\Delta V_2 = \Delta m_2 = \Delta m$. Similarly, since the only external force acting on the fluid in the pipe is gravity, the change in the gravitational potential energy is given by: $\Delta PE = \Delta PE_G = \Delta mgh$. Let the zero point of the gravitational potential energy be on the *ground* level and let the lower end of the pipe be at height y_1 above the ground and let the upper end of the pipe be at height y_2 above the ground. Hence, $h = y_2 - y_1$. Therefore,

$$\Delta PE = \Delta mg(y_2 - y_1) \ .$$

Putting all of the pieces together, the energy equation reads:

$$(P_1 - P_2)\Delta V = \left(\frac{\rho\Delta V v_2^2}{2} + \rho\Delta V g y_2\right) - \left(\frac{\rho\Delta V v_1^2}{2} + \rho\Delta V g y_1\right) \ .$$

After dividing out ΔV and rearranging

$$P_1 + \frac{\rho v_1^2}{2} + \rho g y_1 = P_2 + \frac{\rho v_2}{2} + \rho g y_2 \ .$$

Note that all the quantities on the left hand side of the equality refer to the point R_1 and that all the quantities on the right hand side of the equality refer to the point R_2 . This means that we have a conservation law (also known as Bernoulli's equation) which applies to the fluid. If you look at it more closely you will note that this is just the energy conservation law.[*]

(a) Using Bernoulli's equation we can complete the required calculation and find the quantities we are after. To find the speed at points R_1 and R_2 , use

$$P_1 - P_2 = \frac{\rho}{2}(v_2^2 - v_1^2) + g\rho(y_2 - y_1) \ ,$$

with $P_1 - P_2 = (2.5 - 1.5) \times 10^4 \, \text{Pa} = 1.0 \times 10^4 \, \text{Pa}$. Also, $y_2 - y_1 = h = 0.50\,\text{m}$.

---

[*] In working out this problem we have derived the celebrated Bernoulli's equation. As it is stated it holds for any incompressible, nonviscuous fluid which is moving steady and is not rotating in the gravitational field. You may say that one could have recognized that the Bernoulli's equation applies and just use it. That is true. However we hope that you will appreciate the calculus inspired logic that enables us write down the equation which then holds for any flow (under the conditions described) of the fluid.

Hence $g\rho(y_2 - y_1) = (9.8\,\text{m/s}^2)(1,000\,\text{kg/m}^3)(0.50\,\text{m}) = 0.49 \times 10^4\,\text{Pa}$. Finally, using the fact that the flow rate is constant $A_1v_1 = A_2v_2$, it follows, $v_2 = v_1(A_1/A_2)$. Therefore,

$$v_2^2 - v_1^2 = v_1^2((A_1/A_2)^2 - 1) = v_1^2((r_1/r_2)^4 - 1) = v_1^2((8\,\text{cm}/4\,\text{cm})^4 - 1) = 15v_1^2 \ .$$

Combining all of the pieces yields an equation for v_1,

$$\frac{15(1000\,\text{kg/m}^3}{2}v_1^2 = 1 \times 10^4\,\text{Pa} - 0.49 \times 10^4\,\text{Pa} = 0.51 \times 10^4\,\text{Pa} \ .$$

Hence, $v_1 = 0.82\,\text{m/s}$. Also, $v_2 = (0.82\,\text{m/s})(8\,\text{cm}/4\,\text{cm})^2 = 3.3\,\text{m/s}$.

(b) Finally, we can calculate the flow rate. The fluid flow rate in this case is constant:

$$Q = \frac{\Delta V}{\Delta t} = Av = \text{constant} \ .$$

Therefore, the information at either point can be used. hence,

$$Q = A_1v_1 = \pi(0.08\,\text{m})^2(0.82\,\text{m/s}) = 1.6 \times 10^{-2}\,\text{m}^3/\text{s} \ .$$

As a check you can compute A_2v_2.

EXAMPLE 11.3

(a) Consider a wind-mill with vertical blades of cross-sectional area A. The wind-mill is facing the wind directly. If the wind blows with speed v calculate the maximum possible rate of energy transfer from the wind to the wind-mill blades. (b) If the blade diameter is 80 m and the wind has the speed of 10 m/s what is the power delivered to the wind-mill assuming the overall efficiency of 15%. Consider the case of a wind-mill installed in a windy desert area where the air density is $\rho = 1.16\,\text{kg/m}^3$.

SOLUTION

[Given: wind-mill blade area A facing the wind directly which blows with speed v. Find: (a) the maximum possible power transfer; (b) power delivered when the blade diameter is $2r = 80\,\text{m}$, wind speed is $v = 10\,\text{m/s}$ and the wind-mill efficiency is $\xi = 15\%$; $\rho = 1.16\,\text{kg/m}^3$.]

(a) The geometry of the problem is depicted in Fig. EG-11.3.

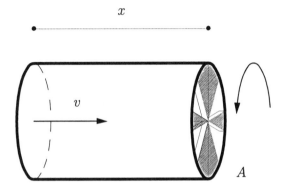

Figure: EG-11.3

Let us observe the motion of the wind-mill for a short time interval Δt. During that time the wind which blows with speed v travels the distance $\Delta x = v\Delta t$. This means that all the air molecules that are set in motion as the wind and that are contained within volume $\Delta V = A\Delta x = Av\Delta t$ will collide with the blades of the wind-mill.*
Assuming that air molecules that collide with the blades of the wind-mill transfer all of their kinetic energy to the wind-mill, the total energy gain of the wind-mill equals to the energy loss of the affected air molecules,

$$\Delta\text{KE} = \frac{1}{2}\rho\Delta V v^2 \,.$$

---

* The assumption used here is not true in a realistic case. A large fraction of the molecules will in fact slip between the blades of the wind-mill. However, it is reasonable to carry out the calculation as described as long as it is understood that this effect can be accounted for through the efficiency of the realistic wind-mill.

From this it is easy to find the rate of energy transfer, *i.e.* the power delivered to the wind-mill.

$$P = \frac{\Delta KE}{\Delta t} = \frac{\rho A v \Delta t v^2}{\Delta t} = \frac{1}{2} A \rho v^3 \ .$$

The calculated power is the maximal possible power that can be delivered to the wind-mill. Note that the power transfer rate depends on the area of the wind-mill blades.

(b) Apply the result of part (a) to the situation described in the problem. The wind-mill is not 100% efficient. This means that only of the fraction of the available power is transferred to the wind-mill. If the wind-mill efficiency is ξ then the true power is $P_{true} = \xi P$. Hence, using the numerical data in the problem yields,

$$P_{true} = (0.15)\frac{1}{2}(\pi(\frac{80\,\text{m}}{2})^2)(1.16\,\text{kg/m}^3)(10\,\text{m/s})^3 = 440\,\text{kW} \ .$$

EXERCISES

EXERCISE 11.1

Consider the dam as described in the Example problem 11.1. Calculate the torque exerted by the water behind the dam on the dam with respect to the axis through the foot of the dam. Also find the effective arm of the force that exerts the torque on the dam. Note that the result explains why the dam is always built such that it is much wider at the bottom than at the top.

EXERCISE 11.2

Show that the variation of the atmospheric pressure with altitude is given by $P = P_0 e^{-\alpha h}$, where P_0 is the atmospheric pressure at some reference level and $\alpha = \rho_0 g / P_0$, where ρ_0 is the air density at the reference level. Assume that the density of air is directly proportional to the pressure.

EXERCISE 11.3

The water flows through a horizontal pipe at a rate of $2\,\text{m}^3/\text{min}$. Determine the velocity of the flow at a point where the diameter of the pipe is $10\,\text{cm}$ and also at another point where it is $5\,\text{cm}$.

EXERCISE 11.4

What is the hydrostatic pressure force on a back of a dam that is $1,200\,\text{m}$ long and when the water is $150\,\text{m}$ deep.

EXERCISE 11.5

The water in some river flows at a rate of $3,200\,\text{m}^3/\text{s}$. What will be the maximum power output of the turbines in the dam if the water falls a vertical distance of $160\,\text{m}$. Assume that the turbines are 90% efficient.

EXERCISE 11.6

A hole is punched in the side of a $20\,\text{cm}$ tall container full of water. If the water is to shoot horizontally as far as possible, how far up the container should the hole be punched? Neglecting friction losses, how far (initially) from the side of the container will the water land?

EXERCISE 11.7

A conical cup of height b and the apex angle 2α rests open-end-down on a smooth flat surface. Cup is filled to a height h with liquid of density ρ. What will be the lifting force on the cup.

EXERCISE 11.8

A tank, $10\,\text{m}^3$ in volume has an intake valve through which the gas is pumped in at a rate of $1\,\text{m}^3/\text{s}$. The gas escapes through an outlet at the same volume rate but at the density existing in the tank which is $1/3$ of the intake density. Find the density of the gas inside the tank.

Find the net change of the density in a tank of volume $0.28\,\mathrm{m}^3$ if a gas is escaping through an outlet, $0.13\,\mathrm{m}$ in diameter at a speed of $305\,\mathrm{m/s}$. The density in the tank at the start of the flow was $16.1\,\mathrm{kg/m}^3$.

EXAMPLE 12.1

A baby is sitting in a baby swing, which is driven by a wind-up motor. When his parent winds the crank, the mechanism allows the baby to swing in a simple harmonic motion (SHM). If the baby's center of mass horizontal displacement, can be described as $x(t) = 0.300\,\mathrm{m}((\cos\frac{\pi\mathrm{rad}}{s})t)$, calculate, using calculus, the baby's velocity and acceleration as a function of time. Also calculate (or identify) the amplitude, frequency, angular frequency and period of the baby's oscillations.

SOLUTION

[Given: $x(t) = C\cos Bt$, where $C = 0.300\,\mathrm{m}$ and $B = \pi\,\mathrm{rad/s}$. Find: $v(t)$, $a(t)$, A, f, ω and T.] This problem is designed to illuminate the interconnections between displacement, velocity, and acceleration for objects moving in SHM. We have

$$v(t) = \frac{dx}{dt} = \frac{d}{dt}(C\cos Bt) = -CB\sin Bt$$

$$a(t) = \frac{dv}{dt} = \frac{d}{dt}(-CB\sin Bt) = -CB^2\cos Bt$$

So, putting in the numbers gives us:

$$v(t) = -CB\sin Bt = -(0.3\,\mathrm{m})(\pi\,\mathrm{s}^{-1})\sin(\pi\,\mathrm{rad/s})t$$

$$v(t) = -(0.942\,\mathrm{m/s})\sin(3.14\,\mathrm{rad/s})t$$

and

$$a(t) = -(0.3\,\mathrm{m})\left(\pi\frac{\mathrm{rad}}{\mathrm{s}}\right)^2\cos\left(\pi\frac{\mathrm{rad}}{\mathrm{s}}\right)t$$

$$a(t) = -(2.96\,\mathrm{m/s^2})\cos(3.14\tfrac{\mathrm{rad}}{\mathrm{s}})t$$

Next, we want to identify (or calculate) some of the characteristics of baby's motion.

- The amplitude is the maximum distance the baby is away from $x = 0$. The amplitude is the number in front of the cosine function in $x(t)$, so $A = C = 0.3\,\text{m}$ or $A = 0.300\,\text{m}$.

- The angular frequency is the number of radians travelled per unit of time. The numbers in front of the t in the cosine function is the angular frequency, ω: $\cos \omega t$. So, $\omega = B = \pi \frac{\text{rad}}{\text{s}}$ or $\omega = \pi \,\text{rad/s}$.

- The frequency is the number of cycles per unit of time. Since we know the angular frequency, we can derive the frequency: $\omega = 2\pi f$, or $f = \frac{\omega}{2\pi} = \frac{\pi \text{rad/s}}{2\pi \text{rad}} = 0.5\,\text{Hz}$. So $f = 0.5\,\text{Hz}$.

- Lastly, we can determine the period from the frequency. The period is the time it takes the baby to complete one cycle.

$$T = \frac{1}{f} = \frac{1}{0.5\,\text{Hz}} = \frac{1}{0.5\,\text{s}^{-1}} = 2\,\text{s}.$$

So, $T = 2\,\text{s}$. It takes two seconds for the baby to go from the farthest back position around to the farthest back position again.

Now that we understand the basics of the motion, let's examine the expressions in a little more detail. *When is the displacement zero?* When the baby is at the bottom of his swing (at his lowest point) his displacement is zero. *When is the velocity zero?* The baby's velocity is zero when he reaches the highest point in his swing. When he reaches his highest point (either to the front or to the back), his displacement is a maximum, but his velocity is zero. This makes sense because, just for an instant, he stops (he was going forward, he stops, then he goes backwards). *What about the acceleration?* The baby's acceleration is zero when his velocity is a maximum because the change in velocity is zero at that point (at the lowest point of the swing). The acceleration is maximum when the velocity is zero because the velocity vector changes direction and that leads to a maximum acceleration. As you can see, the displacement, velocity, and acceleration are related to each other, even in single harmonic motion problems.

EXAMPLE 12.2

EXAMPLE 12.2

An essentially massless spring is connected to the ceiling. A 5.00 g mass is attached to the spring and hangs so that the spring is 2.00 cm longer than its unstretched length. The mass of is then pulled downward so that it is 20 cm below the equilibrium position and released from rest. After 10 seconds, the amplitude of oscillation has decreased to 15 cm, due to the viscous drag of the air. If the motion of the mass can be written in the form: $x(t) = A_0 e^{-bt/2m} \cos \omega_0 t$, calculate an expression for the net force on the mass as a function of time. Note: $\omega_d = \sqrt{\frac{k}{m} - \left(\frac{b}{2m}\right)^2}$ rather than just $\omega = \sqrt{\frac{k}{m}}$ because damping not only decreases the amplitude of oscillation, but also decreases the frequency of oscillation.

SOLUTION

[Given: $x(t) = A_0 e^{-bt/2m} \cos \omega t$, $A_0 = 20$ cm (starting amplitude); $m = 5.0$ g and $\Delta x = 2.0$ cm (used to get k); At $t = 10$ s $A = 15$ cm (used to calculate b.) Find: $F(t)$.]

First, let's determine what forces are acting on the mass: gravity acts leading to $F_w = mg$. The spring acts as: $F_e = kx_e$, and the viscuous drag acts as: $F_d = bv$. So we need to determine the constants in $x(t)$ and we need to calculate $v(t)$. The information we were given can help us determine all the constants. At $t = 10$ s, $A = 15$ cm. Also, $A_0 = 20$ cm and $m = 5.0$ g. So the amplitude, A is given by:

$$A = A_0 e^{-bt/2m} \qquad \text{or} \qquad \ln\left(\frac{A}{A_0}\right) = \frac{-bt}{2m}$$

so

$$b = \frac{-2m}{t} \ln\left(\frac{A}{A_0}\right) = -\frac{2(5.0\text{g})}{10\text{s}} \ln\left(\frac{15\text{ cm}}{20\text{ cm}}\right) = 0.288\,\text{g/s}.$$

Next we need to determine what ω_d is. If we calculate k, then we know every term in ω_d, since $\omega_d = \sqrt{\frac{k}{m} - \left(\frac{b}{2m}\right)^2}$. Now, we can determine k from the static initial conditions. At equilibrium, gravity pulls down on the mass and the spring pulls up on the mass, so $F_G = F_e$, or $mg = k\Delta x$. So, $k = mg/\Delta x$

$$k = \frac{(5 \times 10^{-3}\,\text{kg})(9.8\,\text{m/s}^2)}{2 \times 10^{-2}\,\text{m}} = 2.45\,\text{kg/s}^2 = 2.45 \times 10^3\,\text{g/s}^2$$

And so $\omega = \sqrt{\frac{k}{m} - \left(\frac{b}{2m}\right)^2} = \sqrt{\frac{(2.45\times10^3\text{g/s}^2}{5.0\text{g}} - \left[\frac{0.288\text{g/s}}{2(5.0\text{g})}\right]^2} = 22.1\,\text{rad/s}$. This means that we know all the constants and we can write the expression for $x(t)$:

$$x(t) = (20\,\text{cm})e^{\frac{-(0.288\text{g/s})t}{2(5.0\text{g})}} \cos(22.1\tfrac{\text{rad}}{\text{s}})t \ .$$

Next we need to differentiate x to get v:

$$v(t) = \frac{dx}{dt} = \frac{d}{dt}\left[A_0 e^{-bt/2m} \cos(\omega t)\right]$$

$$= A_0\left(-\frac{b}{2m}\right)e^{-bt/2m}\cos(\omega t) + A_0 e^{-bt/2m}(-\omega)\sin(\omega t)$$

$$= -A_0 e^{-bt/2m}\left[\frac{b}{2m}\cos\omega t + \omega\sin\omega t\right]$$

So now we can get expressions for the forces (where down is positive) $F_G = mg$, $F_e = -kx_e$, and $F_d = bv$. Notice that $x_e = x(t) + 2.0\,\text{cm}$ because gravity stretched the spring 2.0 cm beyond its unstretched length. $x_e =$ length spring is stretched, $x(t) =$ position of mass relative to its equilibrium position. Lastly, we can combine everything and get the net force:

$$F_{total} = F_G + F_e + F_d = mg - kx_e + bv$$

And, plugging in the values, we get:

$$F_{total} = (0.0050\,\text{kg})(9.8\,\text{m/s}^2) - (2.45\,\text{kg/s}^2)\left[(0.20\,\text{m}\,e^{\frac{-(0.288\times10^{-3}\text{g/s})t}{2(0.0050\text{kg})}}\cos\left(22.1\tfrac{\text{rad}}{\text{s}}t\right)\right.$$

$$\left. + 0.020\,\text{m}\right] + (0.288 \times 10^{-3}\,\text{kg/s})\left\{(-0.20\,\text{m}\,e^{\frac{-(0.288\times10^{-3}\text{g/s})t}{2(0.0050\text{kg})}} \ . \right.$$

$$\left.\left[\frac{-(0.288 \times 10^{-3}\text{g/s})t}{2(0.0050\text{kg})}\cos\left(22.1\tfrac{\text{rad}}{\text{s}}t\right) + (22.1\,\text{s}^{-1})\sin\left(22.1\tfrac{\text{rad}}{\text{s}}t\right)\right]\right\}$$

$$F_{total} = 4.5 \times 10^{-2}\,\text{N} - (0.49\,\text{N})e^{-(2.9\times 10^{-2}s^{-1})t} \cos\left(22\tfrac{\text{rad}}{\text{s}}t\right) - 4.5 \times 10^{-2}\,\text{N}$$
$$- \left[(1.7 \times 10^{-6}\,\text{N}) \cos\left(22\tfrac{\text{rad}}{\text{s}}t\right) + (1.3 \times 10^{-3}\,\text{N}) \sin\left(22\tfrac{\text{rad}}{\text{s}}t\right)\right] e^{-(2.9\times 10^{-2}s^{-1})t}$$

$$F_{total} = -\,(0.49\,\text{N})e^{-(2.9\times 10^{-2}s^{-1})t} \cos\left(22\tfrac{\text{rad}}{\text{s}}t\right)$$
$$- \left[(1.7 \times 10^{-6}\,\text{N}) \cos\left(22\tfrac{\text{rad}}{\text{s}}t\right) + (1.3 \times 10^{-3}\,\text{N}) \sin\left(22\tfrac{\text{rad}}{\text{s}}t\right)\right]e^{-(2.6\times 10^{-2}s^{-1})t}$$

EXAMPLE   12.3

A transverse harmonic wave described by

$$y(x,\ t) = (0.0300\,\text{m}) \sin\left[(157\,\text{rad/s})t - (4.83\,\text{rad/m})x\right]$$

propagates on a long wire. Calculate the vertical velocity and the acceleration of the part of the wire at $x = 0.250\,\text{m}$. Calculate (or identify) the period and the wavelength of the wave.

SOLUTION

[Given: $y(x,\ t) = D\sin(Bt - Cx)$, where $D = 0.0300\,\text{m}$, $B = 157\,\text{rad/s}$, and $C = 4.83\,\text{rad/m}$. Find: $\frac{\partial y}{\partial t}$, and $\frac{\partial^2 y}{\partial t^2}$ when $x = 0.250\,\text{m}$, and T and λ.] We first differentiate y with respect to time. We know that the vertical velocity is $\frac{\partial y}{\partial t}$ and the vertical acceleration is $\frac{\partial^2 y}{\partial t^2}$. So,

$$\frac{\partial y}{\partial t} = \frac{\partial}{\partial t}\left(D\sin(Bt - Cx)\right) = DB\cos(Bt - Cx)\,,$$
$$\frac{\partial^2 y}{\partial t^2} = \frac{\partial}{\partial t}\left(\frac{\partial y}{\partial t}\right) = \frac{\partial}{\partial t}\left(DB\cos(Bt - Cx)\right) = -DB^2\sin(Bt - Cx)\,.$$

So,

$$\frac{\partial y}{\partial t} = (0.03\,\text{m})(157\,\text{rad/s})\cos\left[(157\,\text{rad/s})t - (4.83\,\text{rad/m})x\right]$$
$$= (4.71\,\text{m/s})\cos\left[(157\,\text{rad/s})t - (4.83\,\text{rad/m})x\right]$$

and

$$\frac{\partial^2 y}{\partial t^2} = -(0.03\,\text{m})(157\,\text{rad/s})^2 \sin\left[(157\,\text{rad/s})t - (4.83\,\text{rad/m})x\right]$$
$$- (739\,\text{m/s}^2) \sin\left[(157\,\text{rad/s})t - (4.83\,\text{rad/m})x\right]$$

This tells us that at $x = 0.250\,\text{m}$:

$$\frac{\partial y}{\partial t} = (4.71\,\text{m/s}) \cos\left[(157\,\text{rad/s})t - 1.21\,\text{rad}\right]$$

$$\frac{\partial^2 y}{\partial t^2} = -(739\,\text{m/s}^2) \sin\left[(157\,\text{rad/s})t - 1.21\,\text{rad}\right] \ .$$

Next, they ask us to find the period and wavelength of the wave. Recall the expression for a harmonic wave:

$$y(x,t) = A \sin\frac{2\pi}{\lambda}(x - vt)$$

and $v = \lambda f = \frac{\lambda}{T}$, so $C = \frac{2\pi}{\lambda}$ and $B = v\frac{2\pi}{\lambda} = \frac{2\pi}{T}$. Since we know the constants B and C, we can find λ and T.

$$T = \frac{2\pi}{B} = \frac{2\pi\,\text{rad}}{157\,\text{rad/s}} = 0.0400 \text{ seconds}$$

and

$$\lambda = \frac{2\pi}{C} = \frac{2\pi\,\text{rad}}{4.83\,\text{rad/m}} = 1.30\,\text{m}$$

Note: keep in mind that we calculated the vertical velocity and acceleration of the piece of the wire because the wire does not actually *move* in the direction of the propagation of the wave. The propagation velocity of the wave is $v = \frac{\lambda}{T} = \frac{1.3\text{m}}{0.04\text{s}} = 32.5\,\text{m/s}$.

98

EXERCISES

EXERCISE 12.1

A rock is lodged into a car tire. The car is moving at a velocity of 15 m/s. Calculate the height of the rock (relative to the ground) as a function of time. Then, derive an expression, using calculus, for the acceleration of the rock (in the vertical direction) as a function of time. Assume the rock starts on the ground and that the diameter of the tire is 15 inches.

EXERCISE 12.2

A dog's tail oscillates horizontally in simple harmonic motion. If the velocity of the tip of the tail goes as:

$$v(t) = (5.0\,\text{m/s})\sin(\pi\left(22\frac{\text{rad}}{\text{s}}t\right),$$

calculate the amplitude of oscillation of the tip of the dog's tail.

EXERCISE 12.3

Tarzan is swinging on a vine. If he started at a height of 4.5 m above the ground and it takes him 0.75 seconds to reach another branch at the same height, write an expression for his motion as a function of time and using calculus derive an expression for his velocity as a function of time. Assume the two branches are 6.0 m apart, and that he moves in simple harmonic motion (SHM).

EXERCISE 12.4

A pendulum made of a 100 kg mass held up by an almost massless string moves through small angles only. The pendulum is 10.0 m in length and has amplitude of oscillation of 0.200 m. Calculate the kinetic energy of the pendulum as a function of time if the pendulum is started from rest at its maximum height, and only acted on by gravity.

EXERCISE 12.5

A child is swinging on a swing and being pushed by her father. Her father stops pushing and the amplitude of the child's oscillations begins to decrease. If her position as a function of time goes as $x(t) = (1.5\,\mathrm{m})e^{-(0.30\mathrm{s}^{-1}t)}\cos(4.2\frac{\mathrm{rad}}{\mathrm{s}})t$, calculate her velocity as a function of time. If she jumps off at $t = 5.0\,\mathrm{s}$, what is her kinetic energy? (Assume the child's mass is $20\,\mathrm{kg}$.)

EXERCISE 12.6

A $50\,\mathrm{kg}$ bag of sand is held up by a large spring. The spring is stretched $1.2\,\mathrm{m}$ because of the bag of sand. The bag of sand is put into motion so that it oscillates at a frequency of $0.455\,\mathrm{Hz}$. At time $t = 0$, a rip forms in the bag and the sand begins to pour out. As the mass of the sand just starts to decrease, calculate the change of the period of oscillation as a function of mass.

EXERCISE 12.7

A pendulum oscillates at a frequency of $0.18\,\mathrm{Hz}$ when its length is $8.0\,\mathrm{m}$. As we start to slowly pull up on the pendulum's string, we start to shorten the pendulum's length. Calculate the change in the pendulum frequency as a function of length just as we start to shorten the length.

EXERCISE 12.8

Calculate the total kinetic energy of a wave pulse at $t = 0$ described by

$$y(x,t) = (5.0 \times 10^{-2}\,\mathrm{m})\sin\left[(31.4\,\mathrm{rad/m})x - (62.8\,\mathrm{rad/s})t\right] \qquad 0 < x < 0.1\,\mathrm{m}$$

traveling on a rope which has a mass per unit length of $50\,\mathrm{g/m}$.

EXERCISE 12.9

A $55\,\mathrm{kg}$ man jumps from a bridge while connected to a short bungie cord. He falls $12\,\mathrm{m}$ and then bounces back up. He then oscillates at a frequency of $0.30\,\mathrm{Hz}$ and rebounds to a height of 6.0 m. If the unstretched bungie cord is 3.0 m long, calculate his elastic potential energy and kinetic energy 0.70 seconds after he reaches his lowest point for the first time. Assume that there is no damping in this problem.

EXERCISE 12.10

A slinky, suspended from the ceiling, is pulled down so that the tension is $T = (0.60\,\text{N/m})(y + 4.0\,\text{m})$ and the mass per unit length is $\mu = (0.20\,\text{kg/m}^{1/2})(y + 4.0\,\text{m})^{-1/2}$ where y is zero at the lowest point on the slinky and $y = 2.5\,\text{m}$ at the ceiling. A transverse wave pulse, originating at $y = 0\,\text{m}$, is sent into motion up the slinky. Calculate the time it takes for the pulse to reach the ceiling.

EXAMPLE 13.1

A sound wave of frequency f and wavelength λ is propagating through the air. The frequency and the wavelength are related by $v = \lambda f$, where v is the speed of the sound. The displacement of the oscillating part of the media through which the sound moves can be expressed as $s(x,t) = s_0 \cos(\frac{2\pi}{\lambda}x - 2\pi f t)$. Show that the relation $v = \lambda f$ implies that the wave function $s(x,t)$ satisfies the equation

$$\frac{\partial^2 s(x,t)}{\partial x^2} = \frac{1}{v^2}\frac{\partial^2 s(x,t)}{\partial t^2} \ .$$

SOLUTION

[Given: λ, f, $v = \lambda f$; $s(x,t) = s_0 \cos(\frac{2\pi}{\lambda}x - 2\pi f t)$. Show: that the wave amplitude satisfies the wave equation implied by the relation $v = \lambda f$.]

In this problem it is the easiest to go backwards, that is, assume that the the wave function $s(x,t)$ satisfies the equation

$$\frac{\partial^2 s(x,t)}{\partial x^2} = \frac{1}{v^2}\frac{\partial^2 s(x,t)}{\partial t^2} \ ,$$

and then to show that this equation implies the relation $v = \lambda f$.

First calculate the second partial derivatives,

$$\frac{\partial^2 s(x,t)}{\partial x^2} = \frac{\partial}{\partial x}\frac{\partial}{\partial x}\left(s_0 \cos(\tfrac{2\pi}{\lambda}x - 2\pi f t)\right) = -\frac{2\pi}{\lambda}s_0\frac{\partial}{\partial x}\sin(\tfrac{2\pi}{\lambda}x - 2\pi f t)$$

$$= -\left(\frac{2\pi}{\lambda}\right)^2 s_0 \cos(\tfrac{2\pi}{\lambda}x - 2\pi f t) \ .$$

Similarly,

$$\frac{\partial^2 s(x,t)}{\partial t^2} = \frac{\partial}{\partial t}\frac{\partial}{\partial t}\left(s_0 \cos(\frac{2\pi}{\lambda}x - 2\pi f t)\right) = -2\pi f s_0 \frac{\partial}{\partial t}\sin(\frac{2\pi}{\lambda}x - 2\pi f t)$$
$$= -(2\pi f)^2 s_0 \cos(\frac{2\pi}{\lambda}x - 2\pi f t) \ .$$

The wave equation implies:

$$-\left(\frac{2\pi}{\lambda}\right)^2 s_0 \cos(\frac{2\pi}{\lambda}x - 2\pi f t) = -\frac{1}{v^2}(2\pi f)^2 s_0 \cos(\frac{2\pi}{\lambda}x - 2\pi f t) \ .$$

For this equation to hold, the coefficients on both sides must be equal:

$$\frac{4\pi^2}{\lambda^2} = \frac{4\pi^2 f^2}{v^2} \ .$$

In other words, $v^2 = \lambda^2 f^2$. Since all the quantities are positive, upon taking the square root, $v = \lambda f$.

We have just shown that the wave function satisfies a very important equation of physics known as wave equation. You may notice that the wave equation admits many different solutions. As a matter of fact, it is not at all difficult to show that any function that is at least twice differentiable and depends on the position the time in the form $x - vt$, i.e. $f(x,t) = f(x - vt)$ is a solution of the wave equation. Try to show it. This fact has very important implications many of which can be understood only after a more in-depth study of the wave phenomena.

Note also that the formula $v = \lambda f$ can be written as $v = (2\pi f)(\lambda/2\pi)$. In this formula you recognize the angular frequency, $\omega = 2\pi f$. The other factor is usually written as, $k = 2\pi/\lambda$, and is commonly referred to as the *wave-vector*. In terms of the angular frequency and the wave vector the relation reads $v = \omega/k$. In this form the relation is commonly referred to as the dispersion relation. In nature there are many wave-like phenomena that will obey the relation similar (but not identical) to $v = \omega/k$.

EXAMPLE 13.2

Isaac Newton was the first person that derived the formula for the speed of sound, $v = \sqrt{B/\rho}$, where B is the bulk modulus of the air. However, in his derivation he made an erroneous assumption that while the sound propagates through the air, the air obeys Boyle's law. Show that because of this erroneous assumption the speed of the sound calculated by Newton at standard temperature and pressure was by about 15%lower than measured speed of 331 m/s.

SOLUTION

[Given: $v = \sqrt{B/\rho}$, $B = -V dP/dV$, $\rho = 1.29\,\text{kg/m}^3$; $v_{measured} = 331\,\text{m/s}$; STP. Find: v assuming Boyle's law]

Boyle's law states that the pressure of the gas at constant temperature is inversely proportional to the volume, $PV = $ constant. assuming that Boyle's law holds for the sound propagating through the air, the bulk modulus of the air is:

$$B = -V\frac{dP}{dV} = -V\frac{d}{dV}\frac{\text{constant}}{V} = -V\frac{\text{constant}}{-V^2} = \frac{\text{constant}}{V} = P \ .$$

This implies:

$$v = \sqrt{\frac{P}{\rho}} = \sqrt{\frac{1.01 \times 10^5\,\text{Pa}}{1.29\,\text{kg/m}^3}} = 280\,\text{m/s} \ .$$

To find the percent difference, calculate,

$$\frac{|v - v_{measured}|}{v_{measured}} \times 100 = \frac{|280\,\text{m/s} - 331\,\text{m/s}|}{331\,\text{m/s}} \times 100 \approx 15\% \ .$$

Note that by considering the ratio $v_{measured}^2/v^2 = (331\,\text{m/s}/280\,\text{m/s})^2 = 1.14$ one can find the corrected formula for the speed of the sound in the air, $v_s = \sqrt{1.14P/\rho}$.

EXAMPLE 13.3

A sound wave of frequency $f = 512\,\text{Hz}$ is produced by plucking a guitar string strung such that the tension is T_0. What will be the frequency of the sound if the tension is increased by 10%?

SOLUTION

[Given: $f = 512\,\text{Hz}$, when the tension is T_0. Find: f when the tension is increased by 10%.]

Let the mass per unit length of the string be μ. The frequency of the sound is given by $f = (1/2L)\sqrt{T_0/\mu}$, where L is the length of the string. Changing the tension generates the change of frequency, $\Delta f = \Delta T_0/2(2L)\sqrt{T_0\mu}$. The fractional change in the frequency is $\Delta f/f = \Delta T_0/2T_0$. Hence, the fractional frequency change equals to one half of the fractional change in the tension. In our case, $\Delta T_0 = 0.1T_0$. Therefore, $\Delta f/f = 0.1/2 = 0.05$. Finally, $f_{new} = \Delta f + f = 0.05f + f = 1.05f = 1.05(512\,\text{Hz}) = 538\,\text{Hz}$.

EXAMPLE 13.4

Consider the following very simple model of compression waves. Imagine a linear array of massless elastic springs of spring constant K, each attached to point masses m so that a chain is formed, see Fig. EG-13.4.

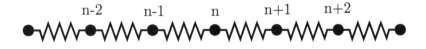

n-2   n-1   n   n+1   n+2

Figure: EG-13.4

For definiteness you may assume that the total number of masses in the chain is N and this means that there are $N - 1$ springs. You may assume that N is a very large number Let us also assume that when the system is in equilibrium, masses are distance a apart from each other. When one of the masses is displaced from the equilibrium position, a longitudinal wave is started in the chain. Since the springs are elastic, each spring acts on the mass attached to its end by an elastic force which equals to the negative of the product of the spring constant and the total spring displacement. Let us say that the mass labeled by n is displaced from the equilibrium by an amount u_n, and that the mass labeled by $n-1$ is displaced from the equilibrium by an amount u_{n-1}. and so on. Knowing these displacements enables us to find the displacement of each spring and therefore to find the force with which springs act on masses attached to them. (a) Show that the acceleration of the mass labeled by n is given by

$$\frac{d^2 u_n}{dt^2} = -\omega_0^2 (2u_n - u_{n-1} - u_{n+1}) ,$$

where $\omega_0^2 = \frac{K}{m}$.

Consider now the limit in which the number of masses and springs, N, becomes infinite and that at the same time the equilibrium distance, a, between masses becomes infinitely small such that the total length of the chain, $L = (N-1)a$, remains unchanged. Assume also that the total mass of the chain, $M = Nm$, remains constant. In other words, the mass m becomes infinitesimally small. In the limit the linear mass density $\mu = M/L = Nm/(N-1)a \approx m/a$ is constant. This limit is commonly known as the continuum limit. You may then assume that the displacement out of the equilibrium becomes a continuum function of the position, $u_n \to u(na) = u(x)$, where, $x = na$, is the position of the $n-th$ mass relative to the beginning of the chain. This means that we can write:

$$u_{n\pm1} = u((n \pm 1)a) = u(x \pm a) \approx u(x) \pm a\frac{\partial u(x)}{\partial x} + \frac{a^2}{2}\frac{\partial^2 u(x)}{\partial x^2} + \cdots .$$

(b) Apply the described limiting procedure to the equation for the acceleration of $n-th$ particle derived in part (a) and show the amplitude $u(x)$ satisfies the wave equation. What is the speed of the wave?

SOLUTION

[Given: an array consisting of a large number N of of masses m and massless elastic springs with spring constant K. Find: (a) the equation for the acceleration of the $n - th$ mass. (b) find the equation of the amplitude $u(x)$ in the continuum limit; speed of the wave.]

(a) This problem is not at all difficult and it relies heavily on the basic concepts of the calculus. Also, it has a deep physical significance because it provides perhaps the simplest possible model for longitudinal waves. The key in finding the answer is in the application of Newton's second law to the motion of each particle in the chain. It is important that you understand the geometry of the array given in Fig. EG-13.4.

Let the displacement of the $n - th$ particle from the equilibrium position be u_n, the displacement of the $n - 1$-st particle be u_{n-1}, and that of the $n + 1$-st particle be u_{n+1}, and so on. For definiteness, let us count the displacement from the equilibrium position to be positive if it is to the right and to be negative if it is to the left. The spring to the left of the $n-th$ particle is stretched (compressed) because the particles n and $n - 1$ have moved. The spring to the right of the n-th particle is stretched (compressed) because the particles n and $n+1$ have moved too. The total stretching (compression) of the spring to the left of the n-th particle is given by $u_n - u_{n-1}$. The stretching (compression) of the spring to the right of the n-th particle is given by $u_{n+1} - u_n$. This means that the spring to the left of the n-th particle acts on the n-th particle with an elastic force, $-K(u_n - u_{n-1})$ and that the spring to the right of the n-th particle acts on the n-th particle with a force, $-K(u_{n+1} - u_n)$. Hence, the total force on the n-th particle is the sum of the forces from the spring just before it and the spring just after it.

Using Newton's second law (remember that the acceleration is the second time derivative of the displacement) the equation of motion of the $n - th$ particle in the chain is:

$$m\frac{\partial^2 u_n}{\partial t^2} = -K(u_n - u_{n-1}) - K(u_{n+1} - U_n) \ .$$

Dividing the expression by the mass and rearranging the terms on the right hand side yields:

$$\frac{\partial^2 u_n}{\partial t^2} = -\omega_0(2u_n - u_{n-1} - u_{n+1}) \ ,$$

where $\omega_0^2 = \frac{K}{m}$.

(b) Let us now perform the continuum limit in the way described in the text of the problem. First, observe that the total length of the mass-spring chain $L = (N-1)a$ remains constant (by assumption) when $N \to \infty$ and $a \to 0$. The total mass of the chain $M = Nm$ will also remain constant (by assumption) when the limit is performed. Note that the assumed limit is more stringent that it needs to be. It would be sufficient to assume that the linear mass density, $\mu = L/M = Nm/(N-1)a = m/(1 - 1/N)a \approx m/a$ remains constant! In any case, this will not affect what follows. In the limit when the number of point masses becomes infinite and distance between them tends to zero it is reasonable to assume that the displacement function u_n becomes a continuous function of the particle position, $x = na$; $u_n = u(na) \to u(x)$: Note also that the particle position is a function of time. However, for the ease of writing the time dependence is suppressed. Therefore, in the limit we can write the displacements of the $n - 1$-st mass as $u_{n-1} = u((n-1)a) \to u(x-a)$, where a is now infinitesimally small. This means that in the limit it is legitimate to approximate,* $u(x-a) \approx u(x) - a\frac{\partial u(x)}{\partial x} + \frac{a^2}{2}\frac{\partial^2 u(x)}{\partial x^2} + \dots$. Similarly, for the displacement of the $n + 1$-st particle we can write, $u_{n+1} = u((n+1)a) \to u(x+a)$. In the limit this approximates as: $u(x+a) \approx u(x) + a\frac{\partial u(x)}{\partial x} + \frac{a^2}{2}\frac{\partial^2 u(x)}{\partial x^2} + \dots$. Let us

---

* see your calculus book if you need to refresh your memory about this type of approximation – you will find the information in the chapter on the expansion of functions into a Taylor series.

now apply these formulas to the acceleration equation derived earlier. Hence,

$$\frac{\partial^2 u(x,t)}{\partial t^2} = -\omega_0^2\left[2u(x,t) - \left(u(x,t) - a\frac{\partial u(x,t)}{\partial x} + \frac{a^2}{2}\frac{\partial^2 u(x,t)}{\partial x^2} + \cdots\right)\right.$$
$$\left. - \left(u(x,t) + a\frac{\partial u(x,t)}{\partial x} + \frac{a^2}{2}\frac{\partial^2 u(x,t)}{\partial x^2} + \cdots\right)\right].$$

Look at the terms on the right hand side. Note that the terms involving first derivatives cancel just like the terms with no derivatives. One is left with: $-a^2\frac{\partial^2 u(x,t)}{\partial x^2} + \cdots$. The omitted terms depend on the parameter a as powers of a higher than the second. As a matter of fact, you can easily check the next omitted term depends on a as a^4! So, in the limit $a \to 0$ the omitted terms will become zero and we do not consider them any further.

How about the term involving the second derivative? The equation looks like a wave equation:

$$\frac{\partial^2 u(x,t)}{\partial t^2} = (a\omega_0)^2\frac{\partial^2 u(x,t)}{\partial x^2} .$$

Now let us determine what happens with $\omega_0^2 a^2$ in the limit. Since $\omega_0^2 = K/M$,

$$\omega_0^2 a^2 = \frac{Ka^2}{m} = \frac{Ka}{m/a} .$$

The ratio in the denominator, m/a, can be written as $Nm/Na = M/L$ which is the linear mass density μ. The numerator term, Ka, can be written as $(K/N)(Na) = K_{eff}L$, where, L, is the total length and, $K_{eff} = K/N$, is the effective spring constant of N springs connected in series. Hence, $K_{eff}L$, is the tension in the system, $T = K_{eff}L$. Therefore, the combination, $\omega_0^2 a^2$, is finite in the limit and

$$\omega_0^2 a^2 = \frac{T}{\mu} .$$

By dimensional analysis it is easy to determine that the quantity $\omega_0 a$ has the dimension of the speed. Hence, write, $v^2 = \omega_0^2 a^2 = T/\mu$. The equation for the displacement in the continuum limit is

$$\frac{\partial^2 u(x,t)}{\partial t^2} = v^2 \frac{\partial^2 u(x,t)}{\partial x^2} .$$

This is the wave equation. The velocity of the longitudinal – compression wave that moves through the system in the continuum limit is $v = \sqrt{T/\mu}$.

The longitudinal wave that moves in this system is a sound wave. This continuum system can be used to model the sound propagation in almost any media. To adapt this model to a system moving in a gas takes very little work. In one of the exercises in this chapter this model is applied to find the speed of the sound wave propagating through the gas.

Let us end this problem by applying the model to find the speed of the sound wave propagating through a solid. Recall that Young modulus of the solid bar is defined as the ratio of the stress and the strain, $Y = \sigma/\epsilon$. The stress is the ratio of the applied force and the area of the cross-section on which the force acts: $\sigma = F/A$. The strain is a measure of the relative elongation (compression) of the bar, the elongation per unit length. If the equilibrium length of the bar is L and the applied stress changes the length by ΔL, then the strain is $\epsilon = \Delta L/L$. Therefore, the applied force can be written as $F = YA\Delta L/L$. Hence, the material, in the elastic region, behaves as a spring with a spring constant $K_{eff} = YA/L$. Let us now assume that the sound wave is propagating through the solid bar. The speed of the sound can be determined from the formula $v^2 = T/\mu$, where $T = K_{eff}L$ and $\mu = M/L$. In this case, $T = YA$. Therefore, $v^2 = YA\mu = Y/(\mu/A)$. Since $\mu/A = \rho$ is the density of the material, the speed of the sound in a solid is given by $v = \sqrt{Y/\rho}$.

EXERCISES

EXERCISE 13.1

In this problem we consider the propagation of the sound through the gas and we will calculate the speed of the sound by the analogy of the system to that of the spring system Example problem 13.4.

Assume that air is confined to a very long cylinder which is closed on one end and has a massless movable piston near the other end. Assume that the pressure of the gas inside the cylinder is P and that the air (gas) density is ρ. The cross-section area of the cylinder is A. The piston can be moved back and forth thus creating the pockets of condensed and rarified gas. These disturbances will propagate through the cylinder as a sound wave. Using this system and its apparent analogy to the spring system considered in Exercise 13.1 show that the speed of the air in the gas can be written as $v = \sqrt{B/\rho}$ where B is the bulk modulus of the gas.

EXERCISE 13.2

A sound wave of frequency f is propagating through the medium in which the speed of the sound is v. The wave function of the oscillating medium can be written as $u(x,t) = u_0 \cos(kx - \omega t)$, where $\omega = 2\pi f$ and $k = \omega/v$. What is the average displacement of the medium? (When speaking of the average we mean the average over the time interval which equals to the the period of the wave motion.) Calculate the average kinetic energy of the medium while the sound is propagating. Calculate the average elastic energy of the medium. [HINT: The average of an arbitrary quantity $G(x,t)$ is defined by $G_{av} = f \int_0^T dt G(x,t)$, where T is the period of the motion and $f = 1/T$ is the frequency.]

EXERCISE 13.3

A sound wave of frequency f_0 is moving through the air and is heard by a stationary listener. If the listener suddenly begins to move toward the source of the sound with constant acceleration a what will be change in the frequency of the sound wave per unit time.

EXERCISE 13.4

(a) A sound of frequency f_0 is moving through an ideal gas at speed v. The velocity of the sound depends on the medium as $v = \sqrt{B/\rho}$, where $B = -V\partial P/\partial V$ is the bulk modulus and ρ is the density of the gas. Remember that sound propagation is an adiabatic process. A temperature of the gas is raised as a function of time like $T = at + T_0$. Find the relative change in the frequency of the sound as a function of time. (b) As a special case consider a sound of frequency $f_0 = 512\,\mathrm{Hz}$ – this is a sound emitted by a simple tuning torque that you may find in the lab and the frequency is measured at STP. What will be the frequency emitted by this tuning torque if the room in which the tuning fork is kept in is heated up from $0^0 C$ to $25^0\,C$ in 5 hours.

EXERCISE 13.5

A sound wave moving through some gaseous medium induces the change in the pressure. Show that the average pressure exerted by the sound wave on the barrier is zero!

EXERCISE 13.6

Show that a wave that travels through the medium exerts zero average pressure on the medium and hence zero average force, and that the energy transfer occurs anyway!

EXAMPLE 14.1

Given a metal which has a coefficient of volume expansion given by: $\beta(T) = (8.0 \times 10^{-6}\,\mathrm{K}^{-1}) + (1.2 \times 10^{-8}\,\mathrm{K}^{-2})T$, calculate the change in volume if it initially had a volume of 0.85ℓ at $T = 20°\,\mathrm{C}$ and the temperature is dropped to $T = 0°\,\mathrm{C}$.

SOLUTION

[Given: $\beta(T) = C + DT$, where $C = 8.0 \times 10^{-6}\,\mathrm{K}^{-1}$ and $D = 1.2 \times 10^{-8}\,\mathrm{K}^{-2}$; $V_0 = 0.85\ell$ at $T_0 = 20°\,\mathrm{C} = 293\,\mathrm{K}$. Find: ΔV if $T_f = 0°\,\mathrm{C} = 273\,\mathrm{K}$.]

We know from equation 14.4 that: $\Delta V = \beta V_0 \Delta T$, so that $\frac{\Delta V}{V} = \beta(T)\Delta T$. Integrating this expression gives:

$$\int_{V_0}^{V_f} \frac{1}{V} dV = \int_{T_0}^{T_f} [C + DT]\, dT$$

$$\ln V \Big|_{V=V_0}^{V=V_f} = \left(CT + \frac{D}{2}T^2\right) \Big|_{T=T_0}^{T=T_f}$$

$$\ln\left(\frac{V_f}{V_0}\right) = \left[\left(CT_f + \frac{D}{2}T_f^2\right) - \left(CT_0 + \frac{D}{2}T_0^2\right)\right]$$

Exponentiating this equation, we get:

$$\frac{V_f}{V_0} = e^{\left[(CT_f + \frac{D}{2}T_f^2) - (CT_0 + \frac{D}{2}T_0^2)\right]}.$$

$$V_f = V_0 e^{[C(T_f - T_0) + \frac{D}{2}(T_f^2 - T_0^2)]}$$

Now, let's actually plug in the values. Remember T_f and T_0 must be in units of

Kelvin ($T_f = 273\,\text{K}$ and $T_i = 293\,\text{K}$).

$$V_f = (0.85\ell)e^{\{(8.0\times10^{-6}\text{K}^{-1})(273\text{K}-293\text{K})+\left(\frac{1.2\times10^{-8}\text{K}^{-2}}{2}\right)[(273\text{K})^2-(293\text{K})^2]\}}$$

$$= (0.85\ell)e^{-2.28\times10^{-4}}$$

This gives us an expression for the final volume. Now we can find ΔV, since $\Delta V = V_f - V_0$.

$$\Delta V = (0.85\ell)e^{-2.28\times10^{-4}} - (0.85\ell)$$

$$= (0.85\ell)[e^{-2.28\times10^{-4}} - 1]$$

$$\Delta V = -1.9 \times 10^{-4}\ell.$$

This answer tells us that the metal decreases its volume by 1.9×10^{-4} liters when its temperature drops from 293K to 273K. This should seem reasonable, since things tend to shrink when they get cooler.

EXAMPLE 14.2

A cylinder is filled with an ideal gas such that its volume is fixed and the number of molecules is also fixed. The container is then heated so that the temperature changes as a function of time:

$$\frac{\partial T}{\partial t} = 0.250\frac{\text{K}}{\text{s}} + (1.00\,\text{K/s}^2)t + (0.500\,\text{K/s}^{3/2})t^{1/2} .$$

Calculate the internal pressure as a function of time, $P(t)$. Assume the initial conditions are: $P = 1.01 \times 10^5\,\text{Pa}$, $V = 3.50\ell$, and $n = 0.145\,\text{moles}$.

SOLUTION

[Given: $\frac{\partial T}{\partial t} = A + Bt + Ct^{1/2}$, where $A = 0.250\,\text{K/s}$, $B = 1.00\,\text{K/s}^2$ and $C = 0.500\,\text{K/s}^{3/2}$; $P_0 = 1.01 \times 10^5\,\text{Pa}$; $V = 3.50\ell$; $n = 0.145\,\text{moles}$. Find: $P(t)$.]

We begin with the ideal gas law, $PV = nRT$, and we know V, n, and R are constants. $P = \frac{nRT}{V}$ can then be differentiated with respect to time to give:

$$\frac{\partial P}{\partial t} = \frac{nR}{V}\frac{\partial T}{\partial t} = \frac{nR}{V}\left[A + Bt + Ct^{1/2}\right] .$$

So, $\int_{P(0)}^{P(t)} dP = \int_{t=0}^{t=t} \frac{nR}{V}[A + Bt + Ct^{1/2}]\, dt$.

$$P(t) - P(0) = \frac{nR}{V}\int_0^t \left[A + Bt + Ct^{1/2}\right] dt$$

$$= \frac{nR}{V}\left[At + \frac{B}{2}t^2 + \frac{2}{3}Ct^{3/2}\right]\Big|_{t=0}^{t=t}$$

$$= \frac{nR}{V}\left[At + \frac{B}{2}t^2 + \frac{2}{3}Ct^{3/2}\right]$$

Therefore, $P(t) = P_0 + \frac{nR}{V}\left[At + \frac{B}{2}t^2 + \frac{2}{3}Ct^{3/2}\right]$.

Note: the units here are confusing. Check that they all work out (remember $1\,\text{J} = 1\,\text{N}\cdot\text{s}$ and $1\,\text{Pa} = 1\,\text{N/m}^2$).

Using the numerical values we have:

$$P(t) = (1.01 \times 10^5\,\text{Pa}) + \frac{(0.145\text{mol})(8.314\text{J/mol K})}{(3.50\ell)(1\text{m}^3/1000\ell)}.$$

$$\left[(0.250\,\text{K/s})t + \frac{(1.00\text{K/s}^2)}{2}t^2 + \frac{2(0.500\text{K/s}^{3/2})}{3}t^{3/2}\right]$$

$$P(t) = 1.01 \times 10^5\,\text{Pa} + (86.1\,\text{Pa/s})t + (172\,\text{Pa/s}^2)t^2 + (115\,\text{Pa/s}^{3/2})t^{3/2}$$

This result tells us that the pressure increases as a function of time, since the temperature increases as a function of time. At $t = 0\,\text{s}$ $P = 1.01 \times 10^5\,\text{Pa}$, and $T = 293\,\text{K}$, but at $t = 10\,\text{s}$, $P = 1.23 \times 10^5\,\text{Pa}$ and $T = 361\,\text{K}$.

Notice that we could have integrated $\frac{\partial T}{\partial t}$ to get $T(t)$ and then substituted into the ideal gas law to get the same expression for $P(t)$. Try it!

EXAMPLE 14.3

Consider the cylindrical container shown below in figure **EG-14.3**. The barrier shown can move freely in order to keep the pressure the same on both sides of it. Compartment one is evacuated at a rate of $\frac{dn_1}{dt} = -\left(0.020\frac{\text{moles}}{\text{s}}\right)$. Assume that the cross-sectional area of the cylinder is $A = 0.50\,\text{m}^2$. The initial volumes are $V_1 = 0.25\,\text{m}^3$ and $V_2 = 0.25\,\text{m}^3$; the initial pressure is $2.0 \times 10^5\,\text{Pa}$. Calculate the initial velocity of the barrier. Assume the cylinder is at a constant temperature of $T = 300\,\text{K}$.

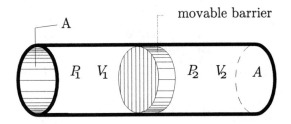

movable barrier

A

P_1 V_1 P_2 V_2 A

Figure: EG-14.3

SOLUTION

[Given: $P_{1_0} = P_{2_0} = 2.0 \times 10^5\,\text{Pa}$, $V_{1_0} = V_{1_0} = 0.25\,\text{m}^3$, $A = 0.50\,\text{m}^2$, $\frac{dn_1}{dt} = -0.020\frac{\text{mol}}{\text{s}}$; $T = 300\,\text{K} = $ constant. Find: $\left.\frac{dx}{dt}\right|_{t=0}$.]

We know that the pressure in each compartment is the same at all times (but the pressure can change with time):

$$P_1(t) = P_2(t) \qquad \text{and} \qquad \frac{\partial P_1}{\partial t} = \frac{\partial P_2}{\partial t} \ .$$

The cylinder remains at a constant temperature, so $\frac{\partial T}{\partial t} = 0$ and therefore,

$$P_2 V_2 = n_2 RT = \text{constant, and} \quad \frac{\partial(P_1 V_1)}{\partial t} = RT\frac{\partial n_1}{\partial t}.$$

116

We also know that the total volume is a constant, so

$$V_1(t) + V_2(t) = \text{constant, and } \frac{\partial V_1}{\partial t} + \frac{\partial V_2}{\partial t} = 0.$$

Lastly, we can get an equation relating V_1 to the position of the barrier, $V_1 = A \cdot x$, so as the volume of compartment one decreases, the barrier moves to the left. This gives:

$$\frac{\partial V_1}{\partial t} = A \frac{\partial x}{\partial t},$$

where $\frac{\partial x}{\partial t}$ is the velocity of the barrier.

Now we must do a lot of algebra to get us an expression for $\left. \frac{\partial V_1}{\partial t} \right|_{t=0}$. First let's write out the equation for the time derivative of $P_1 V_1$:

$$\frac{\partial}{\partial t}[P_1 V_1] = RT \frac{\partial n_1}{\partial t} = \frac{\partial P_1}{\partial t} V_1 + P_1 \frac{\partial V_1}{\partial t}.$$

Now we need an expression for $\frac{\partial P_1}{\partial t}$. Since:

$$\frac{\partial P_1}{\partial t} = \frac{\partial P_2}{\partial t}, \qquad \text{we get:} \qquad RT \frac{\partial n_1}{\partial t} = \frac{\partial P_2}{\partial t} V_1 + P_1 \frac{\partial V_1}{\partial t}$$

Next we need an expression for $\frac{\partial P_2}{\partial t}$. Since:

$$P_2 V_2 = \text{constant}, \qquad \frac{\partial}{\partial t}[P_2 V_2] = 0 \qquad \text{or} \qquad \frac{\partial P_2}{\partial t} V_2 + P_2 \frac{\partial V_2}{\partial t} = 0$$

Therefore, $\frac{\partial P_2}{\partial t} = -\frac{P_2}{V_2} \frac{\partial V_2}{\partial t}$ giving us:

$$RT \frac{\partial n_1}{\partial t} = -\frac{P_2}{V_2} \frac{\partial V_2}{\partial t} V_1 + P_1 \frac{\partial V_1}{\partial t}$$

Lastly, we need an expression for $\frac{\partial V_2}{\partial t}$. Since:

$$\frac{\partial V_1}{\partial t} + \frac{\partial V_2}{\partial t} = 0, \qquad \frac{\partial V_2}{\partial t} = -\frac{\partial V_1}{\partial t}$$

117

and so,

$$RT\frac{\partial n_1}{\partial t} = \frac{P_1}{V_2}V_1\frac{\partial V_1}{\partial t} + P_1\frac{\partial V_1}{\partial t}$$

$$= \left(\frac{P_1V_1}{V_2} + P_1\right)\frac{\partial V_1}{\partial t}$$

$$\frac{\partial V_1}{\partial t} = \frac{RTV_2}{P_1(V_1 + V_2)}\frac{\partial n_1}{\partial t}$$

Substituting into the equation relating $\frac{\partial V_1}{\partial t}$ to $\frac{\partial x}{\partial t}$, we can get $\frac{\partial x}{\partial t}$:

$$A\frac{\partial x}{\partial t} = \frac{RTV_2}{P_1(V_1 + V_2)}\frac{\partial n_1}{\partial t}$$

$$\frac{\partial x}{\partial t}\bigg|_{t=0} = \left(\frac{RTV_2}{AP_1(V_1 + V_2)}\frac{\partial n_1}{\partial t}\right)\bigg|_{t=0} \qquad \text{(put in all values at } t = 0)$$

$$= \frac{(8.314\text{J/mol} \cdot \text{K})(300\text{K})(0.25\text{m}^3)}{(0.50\text{m}^2)(2.0 \times 10^5\,\text{Pa})(0.25\text{m}^3 + 0.25\text{m}^3)}\left(-0.02\frac{\text{mol}}{\text{s}}\right)$$

$$\frac{\partial x}{\partial t}\bigg|_{t=0} = -2.5 \times 10^{-4}\frac{N\text{m}}{\text{m}^2(N/\text{m}^2)\text{s}} = -2.5 \times 10^{-4}\,\text{m/s}$$

So the barrier's initial velocity is 2.5×10^{-4} m/s$-$left.

$$\frac{\partial x}{\partial t}\bigg|_{t=0} = 2.5 \times 10^{-4}\,\text{m/s - left}$$

This problem is a good example of combining information we are given to get the solution. Sometimes many equations are necessary, as in this problem. Be careful with your algebra and check your units at each step, that helps to check your algebra.

EXERCISES

EXERCISE 14.1

A particular material has coefficient of linear expansion given by: $\alpha(t) = 13.00 \times 10^{-6}\,\mathrm{K}^{-1} + (4.100 \times 10^{-8}\,\mathrm{K}^{-3/2})T^{1/2}$. Calculate the length of the object if it initially had a length of $0.3500\,\mathrm{m}$ at $T = 20.00°\,\mathrm{C}$ and the temperature is increased to $T = 100.0°\,\mathrm{C}$.

EXERCISE 14.2

You want to determine the temperature dependence of the coefficient of volume expansion for an unknown metal. You measure the volume of the material at a number of different temperatures and notice that $V(T) = V_0 e^{CT+DT^3}$, where $V_0 = 1.0 \times 10^{-3}\,\mathrm{m}^3$, $C = 25 \times 10^{-6}\,\mathrm{K}^{-1}$ and $D = 1.1 \times 10^{-8}\,\mathrm{K}^{-3}$. Calculate $\beta(T)$.

EXERCISE 14.3

A rod of an unknown metal has a temperature-dependent length, where $L(T) = L_0 e^{CT+DT^{1.7}}$ where $L_0 = 1.5\,\mathrm{m}$, $C = 8.6 \times 10^{-6}\,\mathrm{K}^{-1}$, $D = 4.3 \times 10^{-8}\,\mathrm{K}^{-1.7}$. Calculate the temperature dependence of the coefficient of linear expansion, $\alpha(T)$.

EXERCISE 14.4

A hot-air balloon springs a leak, so that $\frac{\partial n}{\partial t} = -(0.100\,\mathrm{s}^{-1})n$. Assume the balloon starts at $P = 1.00\,\mathrm{atm}$, $T = 40.0°\,\mathrm{C}$, and $n_0 = 200\,\mathrm{moles}$. If the pressure and temperature do not change because of the leak, calculate the expression for the volume of the balloon as a function of time. Assume the hot air behaves as an ideal gas.

EXERCISE 14.5

A cylinder with a piston is heated by placing a flame under it. As the temperature changes, the piston is free to move, in order to keep the pressure constant. If the

119

temperature profile is $T(t) = 293\,\text{K} + (3.0\,\text{K/s})t$, calculate the change in the volume with respect to time, $\frac{\partial V}{\partial t}$. Assume the cylinder is filled with an ideal gas so that the ideal gas law applies, and that there is 1.0 mol of gas at a pressure of $1.01 \times 10^5\,\text{Pa}$.

EXERCISE 14.6

Consider 15.0 mol of an ideal gas. If the pressure of the gas goes as $P(t) = 2.00 \times 10^5\,\text{Pa} - (1.50 \times 10^4\,\frac{\text{Pa}}{\text{s}})t$ and the volume of the gas goes as $V(t) = 7.80\,\text{m}^3 + (0.240\,\text{m}^3/\text{s}^{1/2})t^{1/2}$, find the rate at which the temperature of the gas is changing with time.

EXERCISE 14.7

A balloon holds 7.5 mol of an ideal gas. If the balloon is placed in a chamber where the temperature and pressure can be changed, the volume of the balloon can change. The chamber's temperature varies as $T(t) = 293\,\text{K} - (1.7\,\text{K/s})t$. At the same time the pressure is decreased at a constant rate $\frac{\partial P}{\partial t} = -1.5 \times 10^4\,\text{Pa/s}$. Find the initial rate of change of the volume as a function of time if the initial pressure was $P_0 = 1.01 \times 10^5\,\text{Pa}$, and the initial volume was $0.43\,\text{m}^3$.

EXERCISE 14.8

In an adiabatic process, $PV^{1.4} = C$. Find $\frac{\partial T}{\partial t}$ in terms of $\frac{\partial V}{\partial t}$, P, V, n and R for a fixed amount of ideal gas.

EXERCISE 14.9

Given that the density for water depends on temperature as:

$$\rho(T) = (0.99987 \times 10^6\,\text{g/cm}^3)\left[1.0 + 5.3 \times 10^{-5\circ}\,\text{C}^{-1})T - (6.5 \times 10^{-6\circ}\,\text{C}^{-2})T^2 \right.$$
$$\left. + (1.4 \times 10^{-8\circ}\,\text{C}^{-3})T^3 \right],$$

calculate the temperature dependence of the coefficient of volume expansion, $\beta(T)$, for water.

Exercise 14.10

A thermos bottle is half full with coffee. Assume the initial temperature inside the thermos is $50°\,\text{C}$ and that the temperature decreases as $\frac{\partial T(t)}{\partial t} = -(5.0 \times 10^{-3}\,\text{min}^{-1})(T - 20°\,\text{C})$. Assume also that $V_{bottle} = 1.0\ell$ and $n_{air} = 1.0\,\text{mol}$ (which are constant) and that the coefficient of friction between the thermos lid and the top of the bottle is 0.30. If the mass of the lid is $50\,\text{g}$ and the surface area of the mouth of the thermos bottle is $4.0\,\text{cm}^2$, calculate the minimum initial force needed to open the thermos at $t = 3.0\,\text{hrs}$.

EXAMPLE 15.1

An aluminum (Al) rod, 50 cm long with cross-sectional area of 2.5 cm^2 is partially inserted into a thermally insulated vessel filled with liquid Helium (He) at 4.2 K. The temperature of the rod is 300 K. (a) If one half of the rod length is inserted into the He vessel the He liquid will begin to boil. How many liters of He will boil off by the time the part of the rod immersed in He liquid cools off to the temperature of the liquid He? (Assume that the part of the rod which is outside the vessel does not cool off!) (b) If the temperature of the part of the rod that is outside the He vessel is kept at 300 K what is the boil off rate of the liquid He after the immersed part of the rod has reached the temperature of the liquid He? The thermal conductivity of Al is 31 J/s·cm·K, at the temperature of 4.2 K. The density of Al is 2.7 g/cm^3. The specific heat of Al is 0.21 cal/g^{0}C. The density of He is 0.125 g/cm^3. The latent heat of He is 2.09 J/kg.

SOLUTION

[Given: Aluminum rod, $L_{Al} = 0.50$ m, $A_{Al} = 2.5$ cm^2, $k_{Al} = 31$ J/s·cm·K at the temperature of 4.2 K, $\rho_{Al} = 2.7$ g/cm^3, $T_i = 300$ K; liquid Helium in the vessel is at the temperature $T_f = 4.2$ K; $c_{Al} = 0.21$ cal/g^{0}C, $\rho_{He} = 0.125$ g/cm^3, $L_{He} = 2.09$ J/kg. Find: (a) the volume of boiled-off He until the immersed part of the Al rod reaches the temperature T_f; (b) the boil-off rate of He while the immersed part of the rod is at the temperature T_f and the other part of the rod is kept at T_i.]

Assume that the heat transfer takes place between the Al rod and the liquid He only. This means that during any time interval Δt the amount of heat lost by the rod equals to the amount of heat transferred to the liquid He,

$$Q_{\text{Lost by the } Al \text{ rod}} = Q_{\text{Gained by the liquid } He} .$$

The liquid He in the vessel is kept in equilibrium at its boiling temperature, $T_f = 4.2\,\text{K}$. When the rod is inserted into the vessel, He boils off. The energy transfer balance equation reads:

$$m_{He}L_{He} = m_{Al}c_{Al}\Delta T , \qquad \text{or} \qquad \rho_{He}V_{He}L_{He} = \rho_{Al}V_{Al}c_{Al}\Delta T .$$

This equation means that all of the heat that is transferred from the Al rod to the liquid He goes into converting the liquid He into gaseous He. The energy transfer equation is solved for the volume of the boiled-off volume of He:

$$
\begin{aligned}
V_{He} &= \frac{\rho_{Al}V_{Al}c_{Al}\Delta T}{\rho_{He}L_{He}} \\
&= \frac{(2.7\,\text{g/cm}^3)((2.5\,\text{cm}^2)(25\,\text{cm}))(0.21\,\text{cal/g}^0\text{C})(300\,\text{K} - 4.2\,\text{K})}{(0.125\,\text{g/cm}^3)(2.09\,\text{J/kg})} \\
&= 16.8 l .
\end{aligned}
$$

(b) If the part of the Al rod that is kept outside the liquid He vessel is maintained at the temperature of $300\,\text{K}$, while the other part is already thermalized to the temperature of the liquid He, the heat must be delivered to the rod from the outside to maintain the temperature gradient. This temperature gradient is the reason why the heat is constantly delivered to the liquid He which continues to boil-off.*

$$\frac{dQ_{He}}{dt} = \frac{dQ_{Al}}{dt} = kA\frac{dT}{dx} = kA\frac{\Delta T}{\Delta x} .$$

---

* Notice that the heat flow is from the warmer body toward the cooler body.

Hence,

$$\frac{dQ_{He}}{dt} = \rho_{He}L_{He}\frac{dV_{He}}{dt} = (31 \text{ J/s·cm·K})(2.5 \text{ cm}^2)\frac{300\,\text{K} - 4.2\,\text{K}}{25\,\text{cm}} = 219\,\text{cal/s} \ .$$

Solving for the helium boil-off rate we get

$$\frac{dV_{He}}{dt} = \frac{dQ_{he}/dt}{\rho_{He}L_{He}} = \frac{219\,\text{cal/s}}{(0.125\,\text{g/cm}^3)(2.09\,\text{J/kg})} = 0.351\,\text{l/s} \ .$$

The *He* boil-off rate is $0.351\,\text{l/s}$.

EXAMPLE 15.2

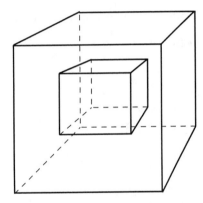

Figure: EG-15.2

A cube shaped container of volume a^3 is kept inside of another cube shaped container of volume b^3 such that the distance between the two containers is the same in all directions. The material separating two containers has thermal conductivity k. Assuming that the walls of the inner cube-shaped container are kept at the temperature T_a and the walls of the outside cube shaped container are kept at the temperature T_b find the rate of the heat transfer. Assume that the heat transfer rate is steady. The geometry is given in Fig. EG-15.2

SOLUTION

[Given: two cube shaped containers of volumes $V_1 = a^3$ and $V_2 = b^3$ respectively, where, $a < b$, are kept such that the smaller container is inside of the bigger one and the distance between them in all directions is the same. The temperatures of the walls are T_a on the wall of the inner container and T_b on the wall of the outer container. The thermal conductivity of the material between them is k. Find: the rate of heat transfer rate, dQ/dt.]

The starting point is the Heat transfer equation

$$\frac{dQ}{dt} = -kA\frac{dT}{dn} ,$$

where dT/dn is the temperature gradient in the direction of the normal pointing from one surface to another. Assume, to make the problem manageable that the edges of the two containers do not matter (in reality this is not likely to be true!). With these assumptions and because of the fact that all sides are identical (bodies are cubes), we can further assume that the heat transfer will take place with the same rate on all 6 pairs of surfaces. Hence, it is sufficient to find the answer for a single pair of faces. Pick any pair of surfaces. Let the direction from the inner surface toward the outer surface be x-direction. Then, $dT/dn = dT/dx$. The heat transfer equation reads

$$\frac{dQ}{dt} = -ka^2\frac{dT}{dx} .$$

Because the heat transfer rate is steady (constant) we can integrate the equation,

$$\int_{T_a}^{T_b} dT = -\frac{dQ/dt}{ka^2} \int_a^b dx .$$

Hence

$$(T_b - T_a) = -\frac{dQ/dt}{ka^2}(b - a) .$$

Solving for the heat transfer rate yields,

$$\frac{dQ}{dt} = ka^2 \frac{T_a - T_b}{b - a} \; .$$

The total heat transfer rate, counting all 6 pairs of surfaces is $6ka^2(T_a - T_b)/(b-a)$.

EXAMPLE 15.3

The passenger section of an airplane can be thought to have a shape of a cylindrical tube. For some small airplane, the tube is 35 m long and has the inner radius of 2.50 m. The exterior of the tubular wall is lined up with a 6 cm thick layer of insulating material of thermal conductivity 4×10^{-5} cal/s·cm·K. If the inside temperature is to be maintained at $25\,^0$C and the outside temperature is $-35\,^0$C, at what rate the heat must be delivered to maintain this temperature difference?

SOLUTION

[Given: cylindrical tube, $L = 35$ m, inner radius $R_i = 2.50$ m lined with $a = 6$ cm thick layer of insulating material of thermal conductivity $k = 4 \times 10^{-5}$ cal/cm·s·K. $T_i = 25\,^0$C, $T_o = -35\,^0$C. Find: dQ/dt assuming it is constant.]

This is the problem of heat transfer with cylindrical geometry. Hence, it is necessary to calculate the heat transfer rate, dQ/dt for the cylindrical geometry. The starting point is the heat transfer equation, $dQ/dt = kAdT/dn$. In case of the cylindrical geometry we can take the direction from the axis of the cylinder as the direction of the normal. This means that $dT/dn = dT/dr$. The inner radius is $R_i = 2.50$ m and the outer radius is $R_o = R_i + a = 2.50$ m + 6 cm = 2.56 cm. The heat is transferred through the insulating material filling the space between the inner and outer radii. The good surfaces to work with, given the cylindrical symmetry of the problem, are cylinders of radius, r, coaxial with the inner and outer cylinders. Therefore,

$R_i \leq r \leq R_o$ and, $A = 2\pi r L$. The heat transfer equation reads:

$$\frac{dQ}{dt} = k2\pi r L \frac{dT}{dr} \ .$$

The equation is integrated to find the heat transfer rate:

$$\int_{T_i}^{T_o} dT = T_o - T_i = \frac{dQ/dt}{2\pi k L} \int_{R_i}^{R_o} \frac{dr}{r} = \frac{dQ/dt}{2\pi k L} \ln \frac{R_o}{R_i}.$$

Therefore, the heat transfer rate is:

$$\frac{dQ}{dt} = 2\pi k L \frac{T_o - T_i}{\ln R_0/R_i} \ .$$

Using numerical values of the parameters in the problem yields,

$$\frac{dQ}{dt} = 2\pi (4 \times 10^{-5} \,\text{cal/cm·s·K})(35\,\text{m}) \frac{-35\,^0\text{C} - 25\,^0\text{C}}{\ln \frac{2.56\,\text{m}}{2.50\,\text{m}}} = -2,200\,\text{cal/s} \ .$$

The heat leaves the airplane cabin at a rate of $2,200\,\text{cal/s}$. If we wish to maintain the temperature difference between the interior and the exterior, then the heat must be delivered to the cabin at the same rate. In other words, the heating system must deliver $2,200\,\text{cal}$ of heat per second. Since $1\,\text{cal} = 4.186\,\text{J}$ the heat power output of the heating system must be $9.2\,\text{kW}$.

EXERCISES

EXERCISE 15.1

A thermos bottle, 30 cm long and having inner radius of 4.0 cm and the outer radius of 4.5 cm with insulating wall of thermal conductivity of $2 \times 10^{-5}\,\text{cal/K cm s}$ is filled with 1 liter of hot coffee of temperature $90\,^0\text{C}$. If the outside wall has the temperature of its surrounding which is $20\,^0\text{C}$, how long it will take for the coffee temperature to fall down to $50\,^0\text{C}$. Assume that the specific heat of coffee is $1\,\text{cal/g}^0\text{C}$.

EXERCISE 15.2

Heat is transferred between two spherical, concentrical containers of radii R_1 and R_2, such that $R_1 < R_2$. The surface of the inner container is at the temperature T_1 and the outer spherical surface is at the temperature T_2. If the wall between them has the thermal conductivity constant k and assuming that the heat is transferred at a constant rate, what is the heat transfer rate?

EXERCISE 15.3

Consider a heat measurement by method of mixtures. Two substances A and B of respective masses m_A and m_B, are brought into a thermal contact while the pressure is kept constant. Suppose that under these conditions, specific heats are $c_A(T)$ and $c_B(T)$, respectively. Assume that before the substances were brought into contact their respective temperatures were T_A and T_B. Some time after the substances are brought into contact they will reach the common final temperature T_f. What is the final temperature T_f. Assume that at the atmospheric pressure the specific heats are temperature independent in this case.

EXERCISE 15.4

Calculate the heat necessary to heat the air in the room of volume $27 \, \mathrm{m}^3$ at a pressure of 1 atmosphere, from the temperature of $0\,^0\mathrm{C}$ to the temperature of $20\,^0\mathrm{C}$. The heat capacity of air is $0.169 \, \mathrm{cal/g} \, \mathrm{K}$. Assume that the air density is $1.29 \, \mathrm{g/l}$.

EXERCISE 15.5

In this problem we will calculate the specific heat of a gas. You know that the specific heat is defined as a rate per unit mass at which the system absorbs the heat. Hence, $c = (dQ/dT)/m$. In the case of the ideal gas the total internal energy of the monoatomic gas is given by $U = 3nRT/2$. The ideal gas also satisfies the equation of state, $PV = nRT$. Show that the specific heat of the system at constant pressure differs from the specific heat at constant volume. Find the relation between specific heats in two cases.

We are accustomed to saying that if we mix equal amounts of the same substance originally at different temperature, the resulting equilibrium temperature of the mixture will be the average of starting temperatures. For example, mixing a pint of water at 95^0 C with a pint of water at 5^0 C will yield two pints of water with temperature of 50^0 C. While this is an important situation it is special in a sense that it is a consequence of the fact that the specific heat of a substance (in our example water) is temperature independent. As an example of a more general situation consider a substance for which in some temperature regime, the specific heat is a function of the temperature of the form $c = kT^3$. If two equal amounts of such a substance, originally at temperatures T_h and T_l are brought into contact, the heat exchange will take place and eventually the equilibrium temperature T_f is reached. Find T_f.

EXERCISE 15.7

A layer of ice of thickness y is formed on a surface of a lake. The air above the ice is at a constant temperature $-T$, $(T > 0)$ and the ice-water interface is at $0\,^0$C. Show that the rate at which the thickness of ice increases in time is given by

$$\frac{dy}{dt} = \frac{kT}{L\rho y} \, ,$$

where, k is the thermal conductivity of ice, L is the latent heat of ice and ρ is the density of ice.

EXAMPLE   16.1

The temperature of $5.0\,\text{mol}$ of an ideal gas changes from $273\,\text{K}$ to $323\,\text{K}$ while the volume changes so that the pressure varies as $P = (1.01 \times 10^5\,\text{Pa}) + (20.0\,\text{Pa/K}^2)T^2$. Calculate the work done by the gas as it undergoes this process.

SOLUTION

[Given: $T_i = 273\,\text{K}$, $T_f = 323\,\text{K}$, $n = 5.0\,\text{mol}$, $P = P_i + aT^2$, where $P_i = 1.01 \times 10^5\,\text{Pa}$ and $a = 20.0\,\text{Pa/K}^2$. Find: W.]

This problem requires us to use the ideal gas law along with the information given to derive an expression for dV, since the work is: $W = \int_{V_i}^{V_f} P\,dV$. Since $PV = nRT$, and P, V and T are changing, the differential form gives us:

$$P\,dV + V\,dP = nR\,dT \ .$$

So, $dV = \frac{1}{P}(nR\,dT - V\,dP)$. This gives $W = \int(nR\,dT - V\,dP)$. Now we need an expression for V and dP in terms of T and dT. Let's start with dP. We know $P = P_i + aT^2$, so

$$dP = 2aT\,dT.$$

Next, we need V in terms of T. We know $PV = nRT$, so $V = \frac{nRT}{P}$ and since

$P = P_i + aT^2$, we get: $V = \frac{nRT}{P_i+aT^2}$. Now, the work becomes:

$$W = \int_{T_i}^{T_f} nR\,dT - \left(\frac{nRT}{P_i + aT^2}\right)(2aT\,dT)$$

$$= \int_{T_i}^{T_f} \left(nR - \frac{2nRaT^2}{P_i + aT^2}\right) dT$$

$$= nR\int_{T_i}^{T_f} 1 - 2\left(1 - \frac{P_i}{P_i + aT^2}\right) dT$$

$$= nR\int_{T_i}^{T_f} \left(-1 + \frac{2P_i}{P_i + aT^2}\right) dT$$

At this point you can either look up the integral, or make the following trig substitution: let $T = \sqrt{\frac{P_i}{a}}\tan\theta$. So, $P_i + aT^2 = P_i + P_i\tan^2\theta = P_i(1+\tan^2\theta) = P_i\cdot\sec^2\theta$ this means that $\frac{2P_i}{P_i+aT^2} = 2\frac{1}{\sec^2\theta} = 2\cos^2\theta$. Now, we need an expression for dT:

$$dT = d\left(\sqrt{\frac{P_i}{a}}\tan\theta\right) = \sqrt{\frac{P_i}{a}}\,d(\tan\theta) = \sqrt{\frac{P_i}{a}}(1 + \tan^2\theta)\,d\theta$$

$$= \sqrt{\frac{P_i}{a}}\sec^2\theta\,d\theta$$

so, the work becomes:

$$W = nR\left[\int -dT + 2\sqrt{\frac{P_i}{a}}\int \cos^2\theta\sec^2\theta\,d\theta\right]$$

$$= nR(-T)\Big|_{T_i}^{T_f} + 2nR\sqrt{\frac{P_i}{a}}\int d\theta$$

$$= nR(T_i - T_f) + 2nR\sqrt{\frac{P_i}{a}}\theta\Big|_{\theta_i}^{\theta_f}$$

and $\theta = \tan^{-1}(T\sqrt{\frac{a}{P_i}})$, so

$$W = nR\left\{ (T_i - T_f) + 2\sqrt{\frac{P_i}{a}}\left[\tan^{-1}\left(T_f\sqrt{\frac{a}{P_i}}\right) - \tan^{-1}\left(T_i\sqrt{\frac{a}{P_i}}\right)\right]\right\}$$

Now, we can substitute in the numbers:

$$W = (5.0\,\text{mol})(8.314\,\text{J/ mol K})\left\{ (273\,\text{K} - 323\,\text{K}) + 2\sqrt{\frac{(1.01 \times 10^5\text{Pa})}{20.0\text{Pa/K}^2}}\right.$$

$$\left.\left[\tan^{-1}\left(323\,\text{K}\sqrt{\frac{20.0\text{Pa/K}^2}{1.01 \times 10^5\text{Pa}}}\right) - \tan^{-1}\left(273\,\text{K}\sqrt{\frac{20.0\text{Pa/K}^2}{1.01 \times 10^5\text{Pa}}}\right)\right]\right\}$$

$$= 41.57\,\text{J/K}\{-50\,\text{K} + 142.13\,\text{K}[77.59 - 75.41]\}$$

$$W = 1.08 \times 10^4\,\text{J}\ .$$

So, the work done by the gas is 10.8 kJ.

This is a very difficult problem, but it illustrates the wide range of uses of the ideal gas law and the work equation:

$$PV = nRT \qquad \text{and} \qquad W = \int_{T_i}^{T_f} P\,dV$$

EXAMPLE 16.2

Calculate the change in the internal energy for 500 mol of hydrogen gas (consider it to be ideal), when the temperature changes as $T = 550\,\text{K} - (10.0\,\text{K}/\ell)V$ as the volume is increased from 25.0ℓ to 35.0ℓ.

SOLUTION

[Given: $n = 500\,\text{mol}$, $c = 14.2\,\text{kJ/kg}\cdot\text{K}$, $m = 504\,\text{g} = 0.504\,\text{kg}$, $T = A - BV$, where $A = 550\,\text{K}$, $B = 10.0\,\text{K}/\ell$, $T_i = 300\,\text{K}$, $T_f = 200\,\text{K}$, $V_i = 25.0\ell$, $V_f = 35.0\ell$. Find: ΔU.]

We know from equation 16.1 that $Q = W + \Delta U$ or $\Delta U = Q - W$, so we need expressions for Q and W. Equation 15.4 says $Q = \int_{T_i}^{T_f} mc\,dT = mcT\big|_{T_i}^{T_f}$

$$Q = (0.504\,\text{kg})(14.2\,\text{kJ/kg}\cdot\text{K})(200\,\text{K} - 300\,\text{K}) = -716\,\text{kJ}.$$

Next, equation 16.5 says $W = \int P\,dV$ and $PV = nRT$. P, V, and T all change.

$$P = \frac{nRT}{V} = nR\left(\frac{A - BV}{V}\right) = nR\left[\frac{A}{V} - B\right]$$

So, we can determine the work done:

$$W = \int_{V_i}^{V_f} nR\left[\frac{A}{V} - B\right] dV = nR\left[A\ln V - BV\right]\Big|_{V_i}^{V_f}$$

$$= nR\left[A\ln\left(\frac{V_f}{V_i}\right) - B(V_f - V_i)\right]$$

$$= (500\,\text{mol})(8.314\,\text{J/mol}\cdot\text{K})\left[550\,\text{K}\ln\left(\frac{35.0\ell}{25.0\ell}\right) - (10.0\,\text{K}/\ell)(35.0\ell - 25.0\ell)\right]$$

$$= 3.54 \times 10^5\,\text{J} = 354\,\text{kJ}$$

So the change in internal energy becomes:

$$\Delta U = Q - W = -716\,\text{kJ} - 354\,\text{kJ} = -1070\,\text{kJ} .$$

So, the internal energy decreases by $1.07 \times 10^3\,\text{kJ}$.

EXAMPLE 16.3

At low temperatures, the specific heat capacity is temperature dependent. For potassium, the specific heat capacity goes as:

$$C = (5.33 \times 10^{-2} \, \text{J/kg} \cdot \text{K}^2)T + (6.59 \times 10^{-2} \, \text{J/kg} \cdot \text{K}^4)T^3$$

Find the heat, Q, required to raise the temperature of 6.0 kg of potassium from 0.10 K to 5.0 K. Calculate, also, the entropy change during this process.

SOLUTION

[Given: $C = \gamma T + AT^3$, where $\gamma = 5.33 \times 10^{-2} \, \text{J/kg} \cdot \text{K}^2$ and $A = 6.59 \times 10^{-2} \, \text{J/kg} \cdot \text{K}^4$; $m = 6.0$ kg, $T_i = 0.10$ K, and $T_f = 5.0$ K. Find: Q and ΔS.]

We know from equation 15.3 that $dQ = mc \, dT$, so integrating gives:

$$Q = \int_{T_i}^{T_f} mc \, dT = m \int_{T_i}^{T_f} (\gamma T + AT^3) \, dT$$

$$= m \left(\frac{\gamma}{2}T^2 + \frac{A}{4}T^4 \right) \Big|_{T_i}^{T_f}$$

$$= m \left(\frac{\gamma}{2}(T_f^2 - T_i^2) + \frac{A}{4}(T_f^4 - T_i^4) \right)$$

Substituting in the known values gives:

$$Q = (6.0 \, \text{kg}) \left\{ \frac{(5.33 \times 10^{-2} \text{J/kg} \cdot \text{K}^2)}{2} [(5.0 \, \text{K})^2 - (0.10 \, \text{K})^2] \right.$$

$$\left. + \frac{6.59 \times 10^{-2} \text{J/kg} \cdot \text{K}^4}{4} [(5.0 \, \text{K})^4 - (0.10 \, \text{K})^4] \right\}$$

$$= 65.8 \, \text{J}$$

So, 66 J of heat are required to increase the temperature of the potassium to 5.0 K.

Next, we need to calculate the change in entropy for this process. Recall from equation 16.20 that $\Delta S = \int \frac{dQ}{T}$. We have an expression for dQ, so

$$\Delta S = \int_{T_i}^{T_f} \frac{(mc\,dT)}{T} = m \int_{T_i}^{T_f} \frac{\gamma T + AT^3}{T}\,dT$$

$$\Delta S = m\left(\gamma T + \frac{A}{3}T^3\right)\Big|_{T_i}^{T_f} = m\left[\gamma(T_f - T_i) + \frac{A}{3}(T_f^3 - T_i^3)\right]$$

Substituting in the known values gives:

$$\Delta S (6.0\,\mathrm{kg})\left[5.33 \times 10^{-2}\,\mathrm{J/kg\cdot K^2}(5.0\,\mathrm{K} - 0.10\,\mathrm{K}) + \frac{6.59 \times 10^{-2}\mathrm{J/kg\cdot K^4}}{3}((5.0\,\mathrm{K}^3 - (0.10\,\mathrm{K})^3)\right]$$
$$= 18.0\,\mathrm{J/K}.$$

So, the entropy of the system *increased* by $18\,\mathrm{J/K}$ during this process.

EXERCISES

EXERCISE   16.1

Calculate the work needed to compress a given amount of oxygen gas from a volume of 2.00ℓ down to 1.50ℓ, assuming the pressure is kept constant (at $P = 1.01\times 10^5\,\mathrm{Pa}$). If the temperature begins at $293\,\mathrm{K}$, what is the final temperature? Use calculus to derive the expression for the work.

EXERCISE   16.2

Calculate the work done by $5.0\,\mathrm{mol}$ of an ideal gas that undergoes isothermal expansion from 5.0ℓ to 15.0ℓ at $T = 300\,\mathrm{K}$.

EXERCISE   16.3

Calculate the total work done in completing the cycle shown:

135

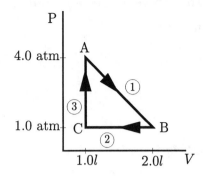

Figure: EX-16.3

Start at $P = 4.0$ atm, $V = 1.0\ell$, go to $P = 1.0$ atm, $V = 2.0\ell$ by a linear change of P with V, then decrease the volume to 1.0ℓ. Lastly, increase the pressure to 4.0 atm to complete the cycle.

EXERCISE   16.4

A diatomic gas, in a chamber with variable volume, undergoes adiabatic expansion, so $PV^{1.4} = $ constant. Calculate the work done by the gas to change its volume from 1.50ℓ to 2.50ℓ. Assume that the gas is ideal, there are 15.0 mol of the gas in the container, and initially the gas is at temperature of 310 K.

EXERCISE   16.5

Consider a 5.50 mol of an ideal gas. If the work done on the gas goes as: $W = 100$ J $+ (50.0 \text{ J/m}^3)V + (10.0 \text{ J/m}^6)V^2$, calculate the temperature as a function of volume.

EXERCISE   16.6

An ideal gas undergoes a process in which the volume increases from 1.50ℓ to 3.00ℓ, and the pressure is varied so that the temperature changes linearly with volume: $T = 293$ K $+ (1.00 \times 10^5 \text{ K/m}^3)V$. If there are 14.7 moles of gas, calculate the work done by the gas.

136

EXERCISE 16.7

Determine the change in internal energy of 5.0 kg of nitrogen gas if the heat capacity goes as:

$$c_v = 6.76\,\text{kJ/kg}\cdot\text{K} + (6.06 \times 10^{-4}\,\text{kJ/kg}\cdot\text{K}^2)T + (1.30 \times 10^{-7}\,\text{kJ/kg}\cdot\text{K}^3)T^2$$

and the gas changes temperature from 250 K to 300 K, but the volume does not change.

EXERCISE 16.8

Consider a Carnot engine with efficiency of $e_c = \frac{T_H - T_L}{T_H}$ where $T_H =$ hot reservoir and $T_L =$ cold reservoir. You are trying to determine how to maximize the efficiency. Do you want to lower the temperature of the cold reservoir or raise the temperature of the hot reservoir? Use calculus to defend your answer.

EXERCISE 16.9

Determine the entropy change of 3.75 kg of acetone when it is cooled from 100° C to 50.0° C.

EXERCISE 16.10

Consider a 35 g sample of a polyethylene substance. If the specific heat capacity varies as

$$c(T) = 2.3\,\text{kJ/kg}\cdot\text{K} + (1.2 \times 10^{-3}\,\text{kJ/kg}\cdot\text{K}^2)T,$$

calculate the change in entropy of the sample when its temperature is changed from 293 K to 393 K.

EXAMPLE   17.1

Consider a charged rod of length L carrying total charge Q uniformly distributed. You may assume that the thickness of the rod is negligible. Calculate the electric field in the space surrounding the rod. Use the result to find the electric field of the infinitely long rod.

SOLUTION

[Given: a rod of length L carrying charge Q. Find: electric field $\mathbf{E}$ in the space surrounding the rod. Consider the limit of the infinitely long rod.]

To calculate the electric field of the rod, imagine that the rod is made out of infinitesimal segments each carrying charge dQ. The electric field of each segment is given by Coulomb's law. The electric field of the entire rod is obtained by adding (integrating) the electric fields of all infinitesimal segments. For the geometry see Fig. EG-17.1

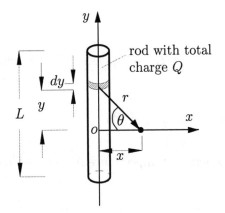

Figure: EG-17.1

Let $\mathbf{r}$ be the vector pointing from the infinitesimal segment of the rod toward the observation point. The electric field of the charged segment is given by $d\mathbf{E} = dQ/\mathbf{r}4\pi\epsilon_0 r^3$. The charge of the infinitesimal segment is given by $dQ = Qdy/L$. We used the assumption that the rod is uniformly charged is used. Note that the coordinate axis are selected such that the rod is along the vertical, y-axis and the x-axis is selected along the line that connects the rod and the observation point. In these coordinates, $\mathbf{r} = x\hat{x} - y\hat{y}$. The electric field of the rod is given by

$$\mathbf{E} = \frac{1}{4\pi\epsilon_0} \int\limits_{rod} \frac{Q}{L} \frac{x\hat{x} - y\hat{y}}{(x^2 + y^2)^{3/2}} dy \ .$$

The electric field has two components which are given in terms of relatively simple integrals,

$$E_x = \frac{Qx/L}{4\pi\epsilon_0} \int\limits_{rod} \frac{dy}{(x^2 + y^2)^{3/2}} \ , \qquad E_y = -\frac{Q/L}{4\pi\epsilon_0} \int\limits_{rod} \frac{y \, dy}{(x^2 + y^2)^{3/2}} \ .$$

The integrals in question are evaluated by a change of variables. First, note that the distance x between the rod and the observation point is constant. So, as one views the rod from the observation point, the good variable appears to be the angle between the horizontal axis and the vector $\mathbf{r}$, see Fig. EG-17.1. Since y is the distance of the infinitesimal segment of the rod from the coordinate origin, the angle at which the infinitesimal element is observed is given by $\tan\theta = y/x$. Hence, $(x^2 + y^2)^{3/2} = x^3(1 + \tan^2\theta)^{3/2} = x^3 \cos^3\theta$. Also, $dy = xd\theta/\cos^2\theta$. it is also useful to denote by θ_1 and θ_2 two angles at which the top and the bottom of the rod are seen from the observation point. They determine the integration endpoints.

The x-component of the electric field (the component pointing away (toward) the rod) reads:

$$E_x = \frac{Qx/L}{4\pi\epsilon_0} \int\limits_{\theta_1}^{\theta_2} \frac{x \cos^3\theta}{x^3 \cos^2\theta} d\theta = \frac{Q/L}{4\pi\epsilon_0 x} \int\limits_{\theta_1}^{\theta_2} \cos\theta \, d\theta$$

$$= \frac{Q/L}{4\pi\epsilon_0 x} (\sin\theta_2 - \sin\theta_1) \ .$$

The integral expression for the y-component of the electric field (the component pointing along the rod) reads

$$E_y = -\frac{Q/L}{4\pi\epsilon_0}\int_{\theta_1}^{\theta_2}\frac{x^2\tan\theta\cos^3\theta}{x^3\cos^2\theta}d\theta = -\frac{Q/L}{4\pi\epsilon_0 x}\int_{\theta_1}^{\theta_2}\sin\theta\,d\theta$$

$$= -\frac{Q/L}{4\pi\epsilon_0 x}(\cos\theta_2 - \cos\theta_1)\ .$$

Hence, the electric field is given by

$$\mathbf{E} = \frac{Q/L}{4\pi\epsilon_0 x}\left[(\sin\theta_2 - \sin\theta_1)\hat{x} - (\cos\theta_2 - \cos\theta_1)\hat{y}\right]\ .$$

To find the limiting case – the electric field of the infinitely long rod, observe that the ratio Q/L is the linear charge density λ of the rod. Also, for the infinitely long rod, the end points are at angles $\theta_2 = \frac{\pi}{2}$ and $\theta_1 = -\frac{\pi}{2}$, respectively. Hence, for the infinitely long rod,

$$\mathbf{E}_\infty = \frac{\lambda}{2\pi\epsilon_0 x}\hat{x}\ ,$$

because the component of the field along the rod vanishes. In the case of the infinite rod the x-component is the radial component and the electric field is in the radial direction.

Note that the result for the electric field of the infinitely long rod can also be obtained by applying Gauss law. The symmetry argument involved while applying the Gauss law is essentially a statement about the vanishing of the component of the electric field along the direction of the rod.

EXAMPLE   17.2

Consider a slab, practically infinite, of uniform volume charge density ρ and thickness d. Calculate the electric field inside and outside of the slab. See Fig. EG-17.2

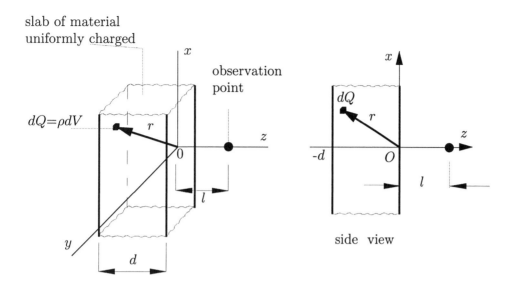

Figure: EG-17.2

SOLUTION

[Given: an infinite slab of uniform charge density ρ and thickness d. Find: electric field everywhere in space.]

To calculate the electric field imagine that the slab is broken into infinitesimal, cube shaped segments with sides, dx, dy and dz, having volume, $dV = dxdydz$, and carrying the charge, $dQ = \rho dV = \rho dxdydz$. The electric field of an infinitesimal segment is given by Coulomb's law,

$$d\mathbf{E} = \frac{1}{4\pi\epsilon_0}\rho\frac{\mathbf{r}}{r^3}dxdydz \ ,$$

where the vector $\mathbf{r}$ points from the infinitesimal segment toward the observation point. The electric field of the slab is obtained by summing (integrating) the contributions of all infinitesimal segments:

$$\mathbf{E} = \int_{slab} d\mathbf{E} = \frac{\rho}{4\pi\epsilon_0}\int_{slab}\frac{\mathbf{r}}{r^3}dxdydz \ .$$

141

The coordinates are selected as shown in Fig. EG-17.2, *i.e.*, the z-axis is perpendicular to the surface of the slab and the observation point is on the z-axis, distance l from the surface of the slab. One surface of the slab coincides with $z = 0$ plane and the other with $z = -d$ plane of the coordinate system. The vector $x\hat{x} + y\hat{y} + z\hat{z}$ is the position of the infinitesimal segment of the slab relative to the coordinate origin; the observation point is at $l\hat{z}$. Hence, the vector pointing from the infinitesimal element toward the observation point is $\mathbf{r} = l\hat{z} - (x\hat{x} + y\hat{y} + z\hat{z}) = -x\hat{x} - y\hat{y} + (l - z)\hat{z}$. The magnitude of vector $\mathbf{r}$ is $r = \sqrt{x^2 + y^2 + (l - z)^2}$. In the selected coordinate system the slab extends from negative to positive infinity in x and y directions and from $-d$ to 0 in the z direction. Therefore, the integral expression for the electric field reads

$$\mathbf{E} = \frac{\rho}{4\pi\epsilon_0} \int_{-\infty}^{\infty} dx \int_{-\infty}^{\infty} dy \int_{-d}^{0} dz \, \frac{-x\hat{x} - y\hat{y} + (l - z)\hat{z}}{[x^2 + y^2 + (l - z)^2]^{3/2}} \, .$$

Using the symmetry argument it is easy to see that integrals representing x and y components of the electric field are zero. For example, the x-component of the electric field is given by

$$E_x = -\frac{\rho}{4\pi\epsilon_0} \int_{-\infty}^{\infty} dx \int_{-\infty}^{\infty} dy \int_{-d}^{0} dz \frac{x}{[(l - z)^2 + x^2 + y^2]^{3/2}} \, .$$

The integral evaluates to zero because the integral over the variable x is taken from $-\infty$ to $+\infty$ and the integrand is an antisymmetric function of x. Similarly, $E_y = 0$. Hence, the electric field is in the z-direction,

$$\mathbf{E} = E_z\hat{z} = \hat{z}\frac{\rho}{4\pi\epsilon_0} \int_{-d}^{0} dz \, (l - z) \int_{-\infty}^{\infty} dx \int_{-\infty}^{\infty} dy \frac{1}{(x^2 + y^2 + (l - z)^2)^{3/2}} \, .$$

The easiest way to evaluate the integral is to perform a change of variables from a cartesian pair of coordinates x and y to radial and angular coordinates by $x =$

$r \cos \phi$, $y = r \sin \phi$. Then, $x^2 + y^2 = r^2$ and $dx dy = r dr d\phi$. Also, the integration bounds are, 0 and ∞ in the radial variable r and 0 and 2π in the angular variable ϕ. Hence,

$$E_z = \frac{\rho}{4\pi\epsilon_0} \int_{-d}^{0} (l - z) \, dz \int_{0}^{\infty} r \, dr \int_{0}^{2\pi} d\phi \frac{1}{[r^2 + (l - z)^2]^{3/2}}$$

$$= \frac{2\pi\rho}{4\pi\epsilon_0} \int_{-d}^{0} (l - z) \, dz \int_{0}^{\infty} \frac{r \, dr}{[r^2 + (l - z)^2]^{3/2}} \cdot$$

The integral over the radial variable is evaluated by a substitution of variable, $u = r^2 + (l - z)^2$. Then, $du = 2r dr$ and the lower bound of the radial integration becomes $(l - z)^2$. To complete the integration over the radial variable use the following result from the integral calculus, $\int du/u^{3/2} = -2/u^{1/2}$. Hence,

$$E_z = \frac{\rho}{2\epsilon_0} \int_{-d}^{0} (l - z) \, dz \int_{(l-z)^2}^{\infty} \frac{1}{2} \frac{du}{u^{3/2}}$$

$$= \frac{\rho}{4\epsilon_0} \int_{-d}^{0} (l - z) \, dz \left[-\frac{2}{u^{1/2}} \right]_{(l-z)^2}^{\infty} = \frac{\rho}{2\epsilon_0} \int_{-d}^{0} (l - z) \frac{1}{\sqrt{(l - z)^2}} dz$$

$$= \frac{\rho}{2\epsilon_0} \int_{-d}^{0} (l - z) \frac{1}{|l - z|} dz \ .$$

The result of the remaining integration depends on whether the observation point is inside or outside the slab.

Let us consider the case when the observation point is outside the slab. Then, $l - z > 0$, which means that $|l - z| = l - z$. The integral over the z variable is simple and the z-component of the electric field reads

$$E_z = \frac{\rho}{2\epsilon_0} \int_{-d}^{0} \frac{l - z}{l - z} \, dz = \frac{(\rho d)}{2\epsilon_0} \ .$$

Hence, the electric field outside the slab is perpendicular to the surface of the slab and it is constant, that is, it is independent of the distance between the slab and the observation point. As a matter of fact, the electric field looks as if it is generated by an infinite charged sheet carrying uniform surface charge density $\sigma = \rho d$. Notice that same result can be found by applying Gauss's law. Try it, it is a good exercise.

Next, consider the case when the observation point is in the interior of the slab. To carry out the integral over the z variable we split the integration region into two parts with the observation point being the divider point. Then,

$$E_z = \frac{\rho}{2\epsilon_0} \left[\int_{-d}^{l} \frac{l-z}{|l-z|}\, dz + \int_{l}^{0} \frac{l-z}{|l-z|}\, dz \right] .$$

In the region from $-d$ to l (to the left of the observation point), $l > z$, and $|l-z| = l - z$. In the region from l to 0 (to the right of the observation point), $l < z$, and $|l-z| = z - l$. Hence, the z-component of the electric field is given by

$$E_z = \frac{\rho}{2\epsilon_0} \left[\int_{-d}^{l} dz - \int_{l}^{0} dz \right] = \frac{\rho}{2\epsilon_0}\left[(l+d) - (0-l)\right] = \frac{\rho}{2\epsilon_0}\left(2l+d\right) .$$

Note that the electric field in the interior of the slab is still perpendicular to the surface of the slab but is not constant. It varies linearly with the distance from the edge of the slab.

The complete expression for the electric field of the slab is given by

$$\mathbf{E} = \begin{cases} -\hat{z}\frac{\rho d}{2\epsilon_0} & \text{for } z \le -d \\ \hat{z}\frac{\rho d}{2\epsilon_0}(2l/d+1) & \text{for } -d \le z \le 0 \\ \hat{z}\frac{\rho d}{2\epsilon_0} & \text{for } z \ge d. \end{cases}$$

Note that the electric field is continuous across the slab. This easily seen if you plot the function $E_z/\frac{\rho d}{2\epsilon_0}$ versus l/d.

EXAMPLE   17.3

A thin rod of length $l = 1$ m carries the total charge $Q = 5\,\mu$ C, such that the linear
charge density depends on the distance from one end of the rod as $\lambda(x) = Ax$. Note
that this means that one end of the rod is not charged at all! Calculate the electric
field of the rod along the axis of the rod and on the side of the charged end. Find
the force that acts on the point charge $q = 1\,\mu$ C placed at a distance $s = 2$ m from
the charged end of the rod. See Fig. EG-17.3

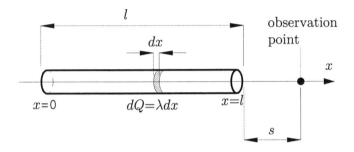

Figure: EG-17.3

SOLUTION

[Given: $L = 1$ m, $Q = 5\,\mu$ C, $\lambda(x) = Ax$; $q = 1\,\mu$ C, $s = 2$ m. Find: $\mathbf{E}$ along the
axis of the rod and on the charged side of the rod; $\mathbf{F}$.]

To find the electric field, imagine that the rod is divided into infinitesimal segments
each carrying charge $dQ = \lambda dx$. The electric field is obtained by integrating the
electric field of a small segments over the length of the rod. Once the electric field
is calculated it is easy to find the force on the charge on the axis of the rod.

Let us first determine the constant A in the expression for linear charge density.
This done by using the fact that the integral of the charge density over the length
of the rod must be equal to the total charge deposited on the rod:

145

$$Q = \int_{rod} \lambda(x)dx = A \int_0^L x\,dx = A\frac{L^2}{2}\,.$$

Hence, $A = 2Q/L^2 = 2(5\,\mu\,\mathrm{C})/(1\,\mathrm{m})^2 = 10^{-5}\,\mathrm{C/m^2}$.

Next, calculate the electric field. The electric field is a vector. Since the observation point is along the axis and to the side of the charged end of the rod and since the charge on the rod is positive, the electric field points in the direction of x-axis, see Fig. EG-17.3. Hence, we can work with scalar values! The electric field of an infinitesimal segment at a point distance s away from the charged end of the rod is given by Coulomb's law formula,

$$dE = \frac{1}{4\pi\epsilon_0}\frac{\lambda(x)dx}{(s+L-x)^2}\,,$$

Hence, the total electric field is an integral of dE.

$$E = \frac{1}{4\pi\epsilon_0}\frac{2Q}{L^2}\int_0^L \frac{x\,dx}{(s+L-x)^2} = \frac{1}{4\pi\epsilon_0}\frac{2Q}{L^2}\int_0^L \frac{x-s-L+s+L}{(s+L-x)^2}dx$$

$$= \frac{1}{4\pi\epsilon_0}\frac{2Q}{L^2}\int_0^L \left[-\frac{dx}{s+L-x} + (s+L)\frac{dx}{(s+L-x)^2}\right]$$

$$\cdot$$

To evaluate the first integral use, $\int dw/w = \ln w$; to evaluate the second integral use, $\int dw/w^2 = -1/w$. Therefore,

$$E = \frac{1}{4\pi\epsilon_0}\frac{2Q}{L^2}\left[\ln(s+L-x)\Big|_0^L + (s+L)\frac{1}{s+L-x}\Big|_0^L\right]$$

$$= \frac{1}{4\pi\epsilon_0}\frac{2Q}{L^2}\left[\ln\frac{s}{s+L} + (s+L)\left(\frac{1}{s} - \frac{1}{s+L}\right)\right]$$

$$= \frac{1}{4\pi\epsilon_0}\frac{2Q}{L^2}\left[-\ln\frac{s+L}{s} + \frac{L}{s}\right]\,.$$

Inserting the numerical values of the parameters in the problem yields,

$$E = (9.0 \times 10^9 \, \text{N m/C}^2) \frac{2(5 \times 10^{-6} \, \text{C})}{(1 \, \text{m})^2} \left[\ln \frac{2 \, \text{m}}{2 \, \text{m} + 1 \, \text{m}} + \frac{1 \, \text{m}}{2 \, \text{m}} \right] = 8.5 \times 10^3 \, \text{N/C} \, .$$

The force on the charge q at the location s is found by $F = qE$: $F = (1 \times 10^{-6} \, \text{C})(8.5 \times 10^3 \, \text{N/C}) = 8.5 \times 10^{-3} \, \text{N}$.

It is a nice exercise to repeat the problem but with the charge placed on the *other*, uncharged, side of the rod. Try it.

EXERCISES

EXERCISE 17.1

A thin strand of plastic is formed into a shape of a semicircle of radius $R = 0.1 \, \text{m}$. It carries a uniform electrical charge density $\lambda = 0.1 \, \mu$ C. A charge $q = 1 \, \text{nC}$ is placed in the in the center of the curvature of the uniformly charged semicircular plastic strand. Find the magnitude and the direction of the electrostatic force acting on the charge q.

EXERCISE 17.2

Consider a uniform electric field **E** oriented in the x-direction. Find the net electric flux through the surface of a cube having side a. Assume that the cube is oriented such that it sides are parallel with coordinate axis!

EXERCISE 17.3

A point charge Q is located a very small distance δ just above the center of the flat face of a hemisphere of radius R. Find, (a) flux through the flat face of the hemisphere, (b) flux through the curved surface of the hemisphere. See Fig. EX-17.3

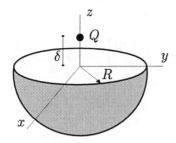

Figure: EX-17.3

EXERCISE 17.4

An electric field is given by $\mathbf{E} = az\hat{x} + bx\hat{z}$, where a and b are constants. Find the flux of the electric field though the face of the flat triangle lying in the x-y plane having its vertices at points, $(0,0)$, $(w,0)$ and (w,h).

EXERCISE 17.5

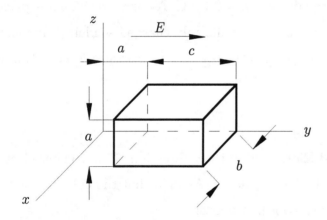

Figure: EX-17.5

A closed surface with dimensions $a = b = 0.4\,\mathrm{m}$ and $c = 0.6\,\mathrm{m}$ is located as shown in the figure EX-17.5. The electric field is nonuniform and given by $\mathbf{E} = (3 + 2x^2)\hat{x}\,\mathrm{N/C}$, where x is measured in meters. (a) Calculate the flux of the electric field through the surface. (b) What amount of electric charge is enclosed by the surface?

Exercise 17.6

In one of the early models of a Hydrogen atom, called Thomson's model, it was assumed that the atom consists of a positive charge q uniformly distributed over the volume of a sphere of radius R. A negative point charge (electron) moves inside the uniform distribution of positive charge. Show that if this model were true the electron would oscillate about the center of the sphere with frequency $f = (1/2\pi)\sqrt{q^2/4\pi\epsilon_0 m R^3}$ where m is the electron mass. What is the numerical value of the oscillating frequency if $R = 0.53 \times 10^{10}$ m and $q = 1.6 \times 10^{-19}$ C. [Note: It is probably needless to say that this model of the atom is incorrect!]

Exercise 17.7

Two parallel planar electrodes (parallel plate capacitor) are separated by a distance of $d = 5$ cm. They are connected to a generator such that the electric field between the plates changes uniformly from 0 to 40 N/C during the time interval $\Delta t = 0.1$ ns. At the moment when the electric field reaches the maximum value, the plates are short circuited so that the electric field immediately falls down back to 0. (a) If an electron was initially at rest next to the cathode when the electric field pulse started, what is the location of the electron just when the pulse ended? (b) How long it is going to take the electron to reach the other electrode (anode)?

Exercise 17.8

A sphere carries a total charge Q which is distributed over the volume according to $\rho = ar^2$. (a) Find the value of the constant a. (b) Find the magnitude and the direction of the electric field inside the sphere. (c) Find the magnitude and the direction of the electric field outside the sphere.

Exercise 17.9

A straight plastic tube of length L and inner radius a and outer radius b is charged uniformly and carries total charge Q. Assume that the tube is very long: $a << L$ and $b << L$. Find the electric field in the space surrounding the tube. [NOTE: because the inner and outer radii of the tube is small compared to the length of the tube you may assume, if necessary, that the tube is infinite]

EXERCISE 17.10

The neutral hydrogen atom in its normal state behaves in some respects like an electric charge distribution which consists of a positive point charge of magnitude $e = 1.6 \times 10^{-19}$ C surrounded by a distribution of negative charge whose density is given by $\rho = -Ce^{-2r/a_0}$, where, $a_0 = 0.53 \times 10^{-8}$ cm, is the so called Bohr's radius and C is some constant needed to make the total value of negative charge equal to $-e$. What is the the electric charge contained within the sphere of radius a_0? What is the electric field strength at distance a_0 from the nucleus?

Chapter 18   ELECTROSTATICS: ENERGY

EXAMPLE 18.1

The electric field, in a particular region of space, is given by: $E_x = (2.34\,\text{V/m}^3)xy$ and $E_y = [(1.17\,\text{V/m}^3)x^2+(4.68\,\text{V/m}^2)y]$. Calculate the potential difference between $(1.00\,\text{m},\ 2.00\,\text{m})$ and $(-2.00\,\text{m},\ -1.00\,\text{m})$.

SOLUTION

[Given: $E_x = Axy$ and $E_y = (Bx^2 + Cy)$, where $A = 2.34\,\text{V/m}^3$, $B = 1.17\,\text{V/m}^3$ and $C = 4.68\,\text{V/m}^2$ $x_i = 1.00\,\text{m}$, $x_f = -2.00\,\text{m}$, $y_i = 2.00\,\text{m}$, $y_f = -1.00\,\text{m}$. Find: ΔV between $(1.00\,\text{m},\ 2.00\,\text{m})$ and $(-2.00\,\text{m},\ -1.00\,\text{m})$.]

We know that $\Delta V = \int_{\ell_i}^{\ell_f} \mathbf{E} \cdot d\ell$. In order to use this equation, we need to pick a path. Let's use the path that gets to the final position in two steps: (1) Go from $(1.00\,\text{m},\ 2.00\,\text{m})$ to $(-2.00\,\text{m},\ 2.00\,\text{m})$ keeping y constant and only changing x, then (2) go from $(-2.00\,\text{m},\ 2.00\,\text{m})$ to $(-2.00\,\text{m},\ -1.00\,\text{m})$ keeping x constant and only changing y. (The choice of path does not change the result, so we chose a path that was easy to integrate.)

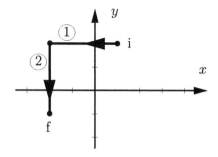

Figure: EG-18.1a

151

So the change in potential becomes:

$$\Delta V = \int_{x=x_i}^{x=x_f} E_x \cdot dx \Bigg|_{y=y_i} + \int_{y=y_i}^{y=y_f} E_y \cdot dy \Bigg|_{x=x_f}$$

or $\Delta V = \int_{x=x_i}^{x=x_f} (Axy_i)\, dx + \int_{y=y_i}^{y=y_f} (Bx^2 + Cy)\, dy.$

$$\Delta V = \frac{A}{2} x^2 y_i \Bigg|_{x_i}^{x_f} + \left(Bx_f^2 y + \frac{C}{2} y^2 \right) \Bigg|_{y_i}^{y_f}$$

$$= \frac{A}{2}(x_f^2 - x_i^2) y_i + Bx_f^2 (y_f - y_i) + \frac{C}{2}(y_f^2 - y_i^2)$$

Plugging in the numbers gives:

$$\Delta V = \frac{(2.34 \text{V/m}^3)}{2} \left[(-2.00\,\text{m})^2 - (1.00\,\text{m})^2 \right] (2.00\,\text{m})$$
$$+ (1.17\,\text{V/m}^3)(-2.00\,\text{m})^2 [-1.00\,\text{m} - (2.00\,\text{m})]$$
$$+ \frac{(4.68\text{V/m}^2)}{2} [(-1.00\,\text{m})^2 - (2.00\,\text{m})^2]$$
$$\Delta V = -14.0\,\text{V}$$

Since the path does not affect the result, let's try a second path. Let's change y first, then change x: So step (1): $x = +1.00\,\text{m}$, $y = 2.00\,\text{m} \rightarrow y = -1.00\,\text{m}$ and

step (2): $y = -1.00\,\text{m}$, $x = +1.00\,\text{m} \rightarrow x = -2.00\,\text{m}$. Then
$\Delta V = \int_{y=y_i}^{y=y_f} E_y dy \Big|_{x=x_i} + \int_{x=x_i}^{x=x_f} E_x dx \Big|_{y=y_f}$.

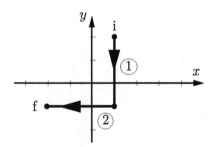

Figure: EG-18.1b

$$\Delta V = \int_{y_i}^{y_f} (Bx_i^2 + Cy)\, dy + \int_{x_i}^{x_f} (Axy_f)\, dx = \left(Bx_i^2 y + \frac{C}{2} y^2 \right)\bigg|_{y_i}^{y_f} + \left(\frac{A}{2} x^2 y_f \right)\bigg|_{x_i}^{x_f}$$

$$= Bx_i^2(y_f - y_i) + \frac{C}{2}[(y_f^2) - (y_i^2)] + \frac{A}{2}[(x_f^2) - (x_i^2)]\, y_f$$

$$= (1.17\,\text{V/m}^3)(1.00\,\text{m})^2(-1.00\,\text{m} - (2.00\,\text{m}))$$

$$+ \frac{4.68\text{V/m}^2}{2}[(-1.00\,\text{m})^2 - (2.00\,\text{m})^2]$$

$$+ \frac{(2.34\text{V/m}^2)}{2}[(-2.00\,\text{m})^2 - (1.00\,\text{m})^2](-1.00\,\text{m})$$

$$\Delta V = -\,14.0\,\text{V}$$

So we see that ΔV is independent of path for this problem. For any physically allowed electric field this will be true.

Challenge: Try this problem again, using a straight-line path. Again, $\Delta V = -14.0\,\text{V}$ should be the answer.

EXAMPLE 18.2

Imagine a bunch of little balls (like b-b's) with mass m and charge $+q$, sitting on a round plastic plate of radius R which has a rim to prevent the balls from rolling away. How will the balls arrange themselves?

Now tilt the plate at an angle of $20°$ with the horizontal. What criterion must the sum of the (gravitational and electrostatic) potential energies satisfy? Find an expression for the electric field produced along the plate by this new configuration of charges.

SOLUTION

[Given: balls with mass m and charge $+q$ and a plate with rim of radius R. Find: the arrangement of charges and with tilt find the electric field.]

First, with the plate flat, the charges will re-distribute so that they are as far apart as possible. To this end, the balls will be equally spaced around the rim of the plate. Notice that the electrostatic potential energy of each of the balls is the same, since $\text{PE} = qV$ and V is the same for each of the balls since they are equally spaced.

Now, with the plate tilted, gravity can influence the position of the balls. The potential energy of each ball is, again, the same, but the potential energy now has a gravitational term as well. So, $\text{PE}_{total} = qV + mgh$, where $h = a \sin 20°$ and $0 < a < 2r$, measured from the point of contact. This means there will be more charges at the lower side of the plate, but the charges will still all be along the rim of the plate.

To derive the expression for the electric field produced, we need an expression for the electric potential all along the plate. From above, $V = \frac{1}{q}[\text{PE}_{total} - mgh]$, where PE_{total} is a constant. $V = \frac{1}{q}[\text{PE}_{total} - mga \sin 20°]$. Now, since we want the electric field along the plate, we want

$$E_a = -\frac{\partial}{\partial a}\left[\frac{\text{PE}_{total}}{q} - \frac{mga \sin 20°}{q}\right] = \frac{mg}{q} \sin 20°$$

$$\Rightarrow E_a \,(\text{with tilt}) = \frac{mg}{q}(\sin 20°)$$

The electric field along the plate is zero before it is tipped because V is the same for all the balls when the plate is flat. Since the gravitational potential energy depends on the height, with the plate tipped, the electric potential, V, is no longer the same for all the balls.

154

EXAMPLE 18.3

Imagine a sphere of radius R and you want to place a total charge of Q evenly over the entire volume. Calculate the total work necessary to bring the charge from infinity and place it on the sphere.

SOLUTION

[Given: R = the radius of the sphere, Q = total charge to be evenly distributed throughout the sphere. Find: W.]

To do this problem, we want to build up the charge in small spherical shells. Imagine that we have already built up charge to a radius a, and we are now adding another shell of thickness da. (See figure **EG-18.3**). So, the work needed to add the charge dq to the shell is: $dW = V\, dq$. Since the charge is distributed evenly over the entire volume, let ρ = charge/unit volume. Then, $dq = (4\pi a^2)\rho\, da$.

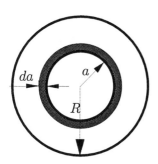

Figure: EG-18.3

Also, we can get an expression for the electric potential due to the already placed charges. Gauss' law says that $E(r) = k\frac{q_{encl}}{r^2}$, where $q_{encl} = \frac{4}{3}\pi\rho a^3$. So, since $V = -\int E \cdot dr$,

$$V(a) = -\int_{\infty}^{a} \frac{kq_{encl}}{r^2}\,dr = \frac{kq_{encl}}{r}\bigg|_{\infty}^{a} = \frac{kq_{encl}}{a} - 0 = \frac{k(\frac{4}{3}\pi a^3)}{a}$$

$$= k\left(\frac{4}{3}\pi a^2\right)$$

So, now we have an expression for dW. Integrating, we get: $\int_0^W dW = \int q\,dV = \int_0^R \left(\rho k\frac{4}{3}\pi a^2\right)\left(4\pi a^2 \rho\,da\right)$.

$$W = k\rho^2\pi^2\frac{16}{15}a^5\bigg|_{a=0}^{a=R} = k\rho^2\pi^2\frac{16}{15}R^5$$

Now, $Q = \rho(\frac{4}{3}\pi R^3)$, so

$$W = kQ^2\left(\frac{9}{15}\right)\frac{1}{R} = \frac{9}{15}k\frac{Q^2}{R} \ .$$

Notice that it requires an amount of work equal to $\frac{9}{15}k\frac{Q^2}{R}$ to place the charges evenly in the spherical volume, but in exercise 18.8 we see that it takes only $\frac{1}{2}k\frac{Q^2}{R}$ to place the charges on the outer surface. That should seem reasonable because the charges are all the same sign and want to be as far apart from each other as possible. In other words, it requires work to move the charges into the center of the sphere from the surface because the charges would prefer to reside on the outer surface.

EXERCISES

EXERCISE 18.1

The electric potential in a particular region of space is given by:

$$V(x,y,z) = (5.75\,\text{V/m}^3)x^3 + (2.70\,\text{V/m}^2)y^2 + (7.37\,\text{V/m})z$$

Determine the x, y and z-components of the electric field. Calculate the magnitude of the electric field at the point $x = 7.00\,\text{m}$, $y = -5.00\,\text{m}$ and $z = 3.00\,\text{m}$.

EXERCISE 18.2

In a particular region in space, the electric potential is described by:

$$V(x, y, z) = (7.15\,\text{V/m}^3)xyz + (2.45\,\text{V/m}^2)yz$$

Determine the x, y and z-components of the electric field. Calculate the magnitude of the electric field at the point $x = 1.00\,\text{m}$, $y = -1.00\,\text{m}$ and $z = 2.00\,\text{m}$.

EXERCISE 18.3

Consider an electron, initially moving at a constant speed v. At time $t = 0.00\,\text{s}$, it begins to feels the effect of an electric potential, given by:

$$V(x, t) = (2.00\,\text{V})\cos\left[(0.300\,\text{m}^{-1})x - (2.00 \times 10^6\,\tfrac{\text{rad}}{\text{s}})t\right].$$

Where must the electron be at $t = 0.00\,\text{s}$ to feel no force by this potential? At what speed must this electron move in order to never feel a force from this potential?

EXERCISE 18.4

A proton moves with an initial velocity of $100\,\text{m/s}$ in the $+x$-direction. At $x = 0.00\,\text{m}$, $y = 0.00\,\text{m}$, it enters a region with a potential of $V(y) = (10.0\,\text{mV/m})(1.00\,\text{m} - \tfrac{1}{2}y)$. Calculate the displacement in the y-direction of the proton when it is at $x = 2.00\,\text{m}$. ($m_{proton} = 1.67 \times 10^{-27}\,\text{kg}$, $q_{proton} = 1.60 \times 10^{-19}\,\text{C}$.)

EXERCISE 18.5

In a given region of space, the electric field is given by:

$$E_x = (6.23\,\text{V/m}^3)x^2$$

Calculate the potential difference between the point $(5.00\,\text{m}, 5.00\,\text{m})$ and the point $(7.00\,\text{m}, 3.00\,\text{m})$.

EXERCISE 18.6

A sphere has charge distributed throughout it such that the charge density goes as $\rho(r) = A(r)^{1/2}$, where $A = 0.0200\,\mathrm{C/m^{7/2}}$. If the sphere has a radius of 2.00 m, calculate the potential at a point 3.00 m from the center of the sphere.

EXERCISE 18.7

Consider a disk of charge having a radius of 10.0 m and a constant charge per unit area, $\sigma = 2.00\,\mu\mathrm{C/m^2}$. Find the potential as a function of distance along the axis perpendicular to the surface of the disk which pierces the center of the disk.

EXERCISE 18.8

You have a sphere of radius R and you want to place a total charge of Q evenly on its surface. Calculate the total work necessary to take the charge from infinity and place in on the surface.

EXERCISE 18.9

The electric potential between two plates of a parallel-plate capacitor is $V = (20.0\,\mathrm{V/m})x$, where $x = 0$ at the left plate of the capacitor, as shown in figure **EX-18.9**.

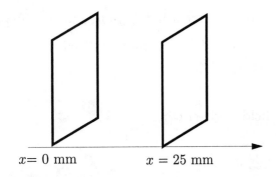

x= 0 mm   x = 25 mm

Figure: EX-18.9

Calculate the electric field between the two plates and draw the equipotential lines

158

at $V = 100\,\text{V}$, $V = 200\,\text{V}$, $V = 300\,\text{V}$ and $V = 400\,\text{V}$. Which plate is positively charged (assume each plate carries the same magnitude of charge).

EXERCISE 18.10

A capacitor, initially with $C = 20.0\,\mu\text{F}$ is charged to a voltage of $10.0\,\text{V}$, and then disconnected from the battery so that the charge remains on the plates. The distance separating the plates is $d = 0.100\,\text{mm}$. A dielectric slab with dielectric constant $\kappa_e = 3.00$ and thickness $0.0700\,\text{mm}$ is then inserted. Calculate the capacitance of the capacitor with the dielectric.

Chapter 19   DIRECT CURRENTS

EXAMPLE 19.1

Two metal spheres connected by a conducting cable and separated by a large distance, and carry total charge $Q = 15\,\mu\text{C}$. The radius on one sphere is $r_1 = 1\,\text{cm}$ and the radius of the other is $r_2 = 3\,\text{cm}$. Imagine now that the sphere of radius r_1 is placed inside of the hollow spherical shell of radius $R = 4\,\text{cm}$ and that the hollow sphere is grounded. Calculate the amount of charge that will flow between the two spheres once the spherical shell surrounding the sphere or radius r_1 is grounded? Neglect the capacitance of the conducting cable connecting the two spheres, see Fig. EG-19.1

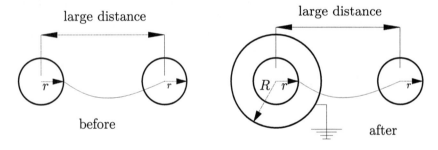

Figure: EG-19.1

SOLUTION

[Given: two spheres, $r_1 = 1\,\text{cm}$, $r_2 = 3\,\text{cm}$, are connected by a conducting cable and are charged such that the total charge distributed between them is $Q = q_1 + q_2 = 15\,\mu\text{C}$. The sphere of radius r_1 is then placed inside of the hollow sphere of radius $R = 4\,\text{cm}$. Find: amount of charge that will flow between the spheres once the hollow shell surrounding the sphere of radius r_1 is grounded?]

The two spheres since they are connected represent an equipotential surface. Because, according to the problem, they are separated by a large distance, the potentials on their surfaces are essentially given by Coulomb's law,

$$V_1 = \frac{1}{4\pi\epsilon_0}\frac{q_1}{r_1} = V_2 = \frac{1}{4\pi\epsilon_0}\frac{q_2}{r_2} .$$

The equality of the potentials on the surfaces of two spheres and the fact that the total charge is $Q = q_1 + q_2$ enables us to calculate charges q_1 and q_2.

$$q_1 = q_2\frac{r_1}{r_2} \quad \text{and} \quad Q = q_1 + q_2 , \quad \text{implies} \quad q_1 = \frac{Qr_1}{r_1 + r_2} , \quad q_2 = \frac{Qr_2}{r_1 + r_2} .$$

Once the sphere of radius r_1 is placed inside of the spherical shell which is then grounded, the configuration of equipotential surfaces changes because the shell which has radius R has potential $V = 0$. This means that the charge distributions will change and that the charge will flow from one sphere to another until the new equilibrium situation is established. Since two spheres of radii r_1 and r_2 remain connected they are still at the same potential. Let us calculate the potential on the surface of the sphere of radius r_1 once it is placed inside of the concentric grounded shell of radius R. The new potential on the surface of the sphere of radius r_1 is

$$V_1' = -\int_R^{r_1} \mathbf{E}\cdot d\mathbf{l} = \frac{q_1'}{4\pi\epsilon_0}\int_R^{r_1}\frac{dl}{r^2} = -\frac{q_1'}{4\pi\epsilon_0}\int_R^{r_1}\frac{dr}{r^2} = \frac{q_1'}{4\pi\epsilon_0}\left(\frac{1}{r_1} - \frac{1}{R}\right) ,$$

and we have used the fact that $dl = -dr$.

The spheres of radii r_1 and r_2, since they are still connected by a long conducting cable, represent an equipotential surface. Hence,

$$V_1' = \frac{q_1'}{4\pi\epsilon_0}\left(\frac{1}{r_1} - \frac{1}{R}\right) = V_2' = \frac{1}{4\pi\epsilon_0}\frac{q_2'}{r_2} .$$

This relation combined with $Q = q_1' + q_2'$ yields expressions for charges q_1' and q_2':

$$q_1' = \frac{Qr_1R}{r_1R + r_2(R - r_1)} , \quad q_2' = \frac{Qr_2(R - r_1)}{r_1R + r_2(R - r_1)} .$$

Next, calculate the amount of charge transported between the two spheres. From

$Q = q_1 + q_2 = q_1' + q_2'$, it follows, $\Delta q = q_1 - q_1' = q_2' - q_2$. Therefore,

$$\Delta q = Q \left(\frac{r_1}{r_1 + r_2} - \frac{r_1 R}{r_1 R + r_2(R - r_1)} \right) = \frac{Q r_1^2 r_2}{(r_1 + r_2)[R(r_1 + r_2) - r_1 r_2]} = 0.86 \, \mu\text{C} \ .$$

EXAMPLE 19.2

A coaxial cable can be visualized as a conducting rod surrounded by a layer of high resistance material of resistivity ρ which is lined by a conducting cylindrical shell. For definiteness, assume that the the radius of the inner rod is r_1 and that the radius of outer cylindrical shell is r_2 and that the outer conductor has negligible thickness. Also assume the idealized situation such that the resistivity of the core rod and the outer conductor is negligible.

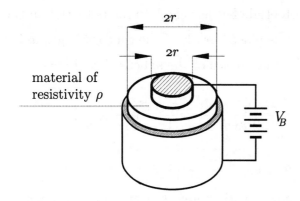

Figure: EG-19.2

If the cable is connected to the battery of voltage V_B, such that the outer conducting surface is connected to the negative pole of the battery and the inner core rod is connected to the positive pole of the battery, (a) find the total current flowing through the high resistance material, (b) the current density $\mathbf{J}$ and the electric field $\mathbf{E}$ at some point P between the core rod and the outer conductor and (c) the resistance of the segment of the cable of length l. See Fig. EG-19.2

SOLUTION

[Given: coaxial cable with inner conductor of radius r_1 and outer cylindrical shell conductor of radius r_2 and negligible thickness; the space between the two conductors is filled with high resistance material of resistivity ρ. The cable is connected to the battery of voltage V_B. Find: (a) I flowing through the high resistance material, (b) $\mathbf{J}$ and $\mathbf{E}$ in the space between the inner and outer conductors and (c) the resistivity of the segment of the cable of length l.]

(a) When the cable is connected to the battery the current I flows radially between the rod and the cylinder. The current density and the electric field are both radial and point away from the axis of the cable, $\mathbf{J} = J\hat{r}$ and $\mathbf{E} = E\hat{r}$. Assuming that the material filling the space between the inner and the outer conductors obeys Ohm's law, then $\mathbf{E} = \rho\mathbf{J}$. Furthermore, if the radial current is uniform, then the current density is given by $J = I/2\pi rl$ where l is a length of some portion of the cable. Therefore, $E = \rho I/2\pi rl$. Hence, in the space between the inner core and the outer conductor, the current density and the electric field decrease inversely proportional to the distance from the central axis. The fact that the electric field fills the space between the inner and the outer conductor implies that there is a potential difference between the inner and the outer surface. The potential difference between the equipotential surfaces, distance dr apart is:

$$dV = -\mathbf{E} \cdot d\mathbf{l} = -Edr = -\frac{\rho I}{2\pi l}\frac{dr}{r} \; .$$

The total electric potential is found by integrating dV along some path leading from the inner toward the outer surface. In the case at hand, the potential difference between two conductors must be equal to the voltage of the battery that the cable is connected to. From the fact that the outer conductor is connected to the negative pole of the battery and that the inner conductor is connected to the positive pole of the battery, the potential difference between the inner and the outer conducting shell is

$$-V_B = \int_{r_1}^{r_2} dV = -\frac{\rho I}{2\pi l} \int_{r_1}^{r_2} \frac{dr}{r} = -\frac{\rho I}{2\pi L} \ln \frac{r_2}{r_1} \ .$$

This equation is solved for the current I, yielding

$$I = \frac{2\pi l}{\rho \ln r_2/r_1} V_B \ .$$

(b) From the expression for the current it is easy to determine the current density and the electric field in the space between the inner and the outer conductors:

$$J = \frac{I}{2\pi rl} - \frac{V_B}{r\rho \ln r_2/r_1} \ ,$$

and

$$E = \rho J = \frac{V_B}{r \ln r_2/r_1} \ .$$

(c) Since the material filling the space between the inner and outer conductors is Ohmic, the current and the driving voltage are related by $V_B = IR$. Hence, the resistance of the segment of the cable of length l is

$$R = \frac{V_B}{I} = \frac{\rho}{2\pi l} \ln \frac{r_2}{r_1} \ .$$

EXAMPLE   19.3

The current density across the cylindrical conductor of radius R varies according to $J = J_0(1 - r/R)$ where r is the distance from the axis. Calculate the current flowing through the conductor.

SOLUTION

[Given: cylindrical conductor of radius R, with current density $J = J_0(1 - r/R)$. Find: I.]

The current density is given by $J = J_0(1 - r/R)$. This means that the current density has a maximum at the center of the conductor, $r = 0$ and that it falls off linearly toward the surface; on the surface it vanishes. The current flowing through the conductor is given by an integral of the current density over the cross-section of the conductor,

$$I = \int_{area} J dA = \int_0^R J_0 \left(1 - \frac{r}{R}\right) 2\pi r dr = 2\pi J_0 \left(\frac{R^2}{2} - \frac{R^3}{3R}\right) = \frac{J_0}{3}\pi R^2 \ .$$

Hence, the current through the conductor is given as one third of the product of the maximum current density J_0 and the area of the cross-section, πR^2, of the conductor. Notice that if the current density was uniform the current flowing through the conductor would be just $J_0\pi R^2$.

EXERCISES

EXERCISE 19.1

The electrical charge flows away from the source according to the formula,

$$q(t) = (2.00\,\mathrm{C/s^3})t^3 + (6.00\,\mathrm{C/s^2})t^2 \ .$$

Find the electrical current I.

EXERCISE 19.2

The current density across the cross-section of the conductor varies as $J = J_0 r/R$. Hence, the current density is minimal (vanishes) at the center of the conductor and it takes the maximal value on the surface of the conductor. Calculate the total current flowing through the conductor.

EXERCISE 19.3

A water flows through a garden hose at a rate of $500\,\text{cm}^3/\text{s}$. To what current of positive charge does this correspond? (Note: obviously, there is no net current flow because each water molecule is electrically neutral. However, since the molecule consists of charged particles - negatively charged electrons and positively charged hydrogen and oxygen nuclei, there is current flow of both, positive and negative charges which compensate each others current!)

EXERCISE 19.4

Two graphite rods of equal length l but different cross sections, having radii a and b respectively, are connected by a conical graphite section of length l that changes its radius linearly from value a on one end to value b on the other. For definiteness assume that $a < b$. Find the electrical resistance of the conical connector. See Fig. EX-19.4.

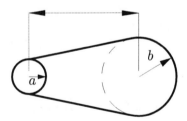

Figure: EX-19.4

EXERCISE 19.5

Consider a block of material of resistivity $\rho = 3.0 \times 10^{-2}\,\Omega$ m and cross-sectional area $A = 10\,\text{cm}^2$. If the electrostatic potential inside of the the block of the material varies with distance as $V = V_0 \sin kx$, where $k = 1\,\text{m}^{-1}$ and $V_0 = 150\,\text{V}$. what is the current flowing through the block? Assume that the material in question obeys Ohm's law.

EXERCISE 19.6

There is a part of Earth's atmosphere which is called the ionosphere. It is a layer of air (actually several layers with somewhat differing properties) that contains many ionized air molecules (mainly N_2 and O_2 molecules). The ionization of molecules occurs mainly by absorption of UV radiation emitted by the Sun. The ion and the free electron density is the greatest at about 200 to 400 km above the surface of the Earth. Because there is an equal amount of positive and negative charge in the ionosphere there is no net electrostatic field. However, because there are ionized molecules and free electrons, there are local electric fields which act on the charges. Such a system is called plasma.

Consider the following simplified model of the plasma. Imagine two walls separated by distance x having excess charge Q and $-Q$ respectively. Consider an electron moving in this plasma and consider its motion under the influence of the plasma electric field only. Calculate the instantaneous electron current in the plasma. You will find out that the electron current oscillates with frequency called the plasma frequency. It is known that the typical daytime frequency of the atmospheric plasma is between 10 and 30 MHz. Use this information and the result of your model calculation to estimate the density of free electrons in Earth's ionosphere. [HINT: consider the force acting on the electron in the plasma due to the electrons that have been displaced such that charged walls are formed! Use Newton's second law to establish the relation between the acceleration of the electron and the electrical force acting on the electron.

EXERCISE 19.7

The electric charge in the system is given by the formula, $Q(t) = Q_0(1 - e^{-t/\tau})$. Find the current in this system. What is the significance of the constant τ.

The current density flowing through the cylindrical rod of radius R is given by $J = J_0(1 - e^{-r/\delta})$. What is the current flowing through the rod?

EXAMPLE 20.1

Consider a capacitor in a series with a resistor, a battery, and a switch, as shown in figure **EG-20.1**.

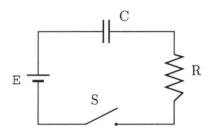

```
Figure:    EG-20.1
```

The capacitor has a capacitance of $17.5\,\mu F$ and initially is completely uncharged. The resistor has a resistance of $250\,\Omega$ and the battery has an emf of $15.0\,V$. Calculate the charge on the capacitor $5.00\,ms$ after the switch is closed.

SOLUTION

[Given: $C = 17.5\,\mu F = 17.5 \times 10^{-6}\,F$, $R = 250\,\Omega$; $Q_i = 0$; $\varepsilon = 15\,V$. Find: $Q(t = 5.00\,ms = 5.00 \times 10^{-3}\,s)$.]

Let's call the time $t = 0\,s$, when the switch is closed. When the switch is closed, the voltage gain at the battery equals the voltage drop across the capacitor and resistor, so:

$$\varepsilon = \frac{Q}{C} + IR, \quad \text{and} \quad I = \frac{dQ}{dt}$$

$$\Rightarrow \varepsilon = \frac{Q}{C} + \frac{dQ}{dt}R$$

To solve this, we solve the equation without the constant, then put the constant in by hand. So we solve: $\frac{Q}{C} = -\frac{dQ}{dt} R$ or

$$\int -\frac{1}{RC} dt = \int \frac{dQ}{Q}$$

$$-\frac{t}{RC} + k = \ln Q$$

$$\Rightarrow Q = e^{(-t/RC+k)}$$

$$= b e^{-t/RC}$$

where b is a constant determined by the boundary conditions. So, $Q = b e^{-t/RC} + C \cdot \varepsilon$ should solve the equation. Let's check.

$$\varepsilon = \frac{Q}{C} + \frac{dQ}{dt} R$$

$$\varepsilon = \left(\frac{b}{C} e^{-t/RC} + \varepsilon \right) + \frac{-b}{RC} e^{-t/RC} R$$

Now, using the initial conditions, we can solve for the constant b using our expression for Q: $Q = 0$ at $t = 0$.

$$0 = b e^{-0/RC} + C \cdot \varepsilon = b + C \cdot \varepsilon \qquad \text{so} \qquad b = -C \cdot \varepsilon$$

and $Q = C \cdot \varepsilon (1 - e^{-t/RC})$. Now, we can substitute in the values we know:

$$Q = (15.0 \, \text{V})(17.5 \times 10^{-6} \, \text{F})(1 - e^{-(-5.00 \times 10^{-3} \text{s})/(17.5 \times 10^{-6} \text{F})(250 \, \Omega)})$$

$$Q(t = 5.00 \, \text{ms}) = 1.79 \times 10^{-4} \, \text{C} = 17.9 \, \text{mC}$$

So, after $5.00 \, \text{ms}$, "capacitor has a charge of $17.9 \, \text{mC}$.

EXAMPLE 20.2

Consider the circuit shown in figure **EG-20.2**. Switch $S1$ is closed for exactly 0.45 seconds, then exactly when switch $S1$ is opened, switch $S2$ is closed. Calculate the time it takes for the charge on the capacitor to decrease from its maximum value to $0.100 \, \text{mC}$.

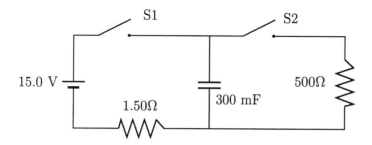

Figure: EG-20.2

SOLUTION

[Given: $\varepsilon = 15.0\,\text{V}$; $R_1 = 1.50\,\Omega$; $C = 300 \times 10^{-3}\,\text{F}$; $R_2 = 500\,\Omega$. Find: t_2.]

First, we need to figure out the charge built up on the capacitor with $S1$ closed. Then, we can figure out how long it takes to dissipate the charge to $0.100\,\mu\text{C}$ and get t_2. So, let's first consider the left loop. When the switch $S1$ is closed, the voltage equation becomes:

$$\varepsilon = \frac{Q}{C} + IR = \frac{Q}{C} + \frac{dQ}{dt}R$$

The solution to this equation is $Q = C\varepsilon(1 - e^{-t/RC})$, see Example 20.1. Let's check:

$$\varepsilon = \frac{C\varepsilon}{C}\left(1 - e^{-t/RC}\right) + \left(\frac{C\varepsilon}{RC}e^{-t/RC}\right)R$$

which is correct. So, using this expression, we can determine the charge on the capacitor at $t = 0.45\,\text{s}$:

$$Q = (300 \times 10^{-3}\,\text{F})(15\,\text{V})\left[1 - e^{-(0.45\text{s})/(1.50\Omega)(300\times 10^{-3}\text{F})}\right]$$
$$= 2.48\,\text{C}$$

So, now we know the initial charge on the capacitor when switch $S2$ is closed. If switch $S2$ is closed at $t = 0$, then the time, t, we determine will tell us how long

171

it takes the charge to decrease to 0.100 mC. Examining the right loop, the voltage equation is: $\frac{Q}{C} = IR = \frac{-dQ}{dt} R$.

$$\int_{Q_0}^{Q} \frac{dQ}{Q} = -\int_{0}^{t} \frac{1}{RC} dt \qquad \ln Q \Big|_{Q_0}^{Q} = -\frac{t}{RC} \Big|_{0}^{t}$$

so $\ln \frac{Q}{Q_0} = -\frac{t}{RC}$.

Now, substituting in the parameters, $Q = 0.100 \times 10^{-3}$ C, $Q_0 = 2.84$ C so

$$t = -R_2 C \ln \frac{Q}{Q_0} = -(500\,\Omega)(300 \times 10^{-3}\,\text{F}) \ln \left(\frac{0.100 \times 10^{-3}\,\text{C}}{2.48\,\text{C}} \right)$$

$$t = 1.54 \times 10^3\,\text{s} = 25.6\,\text{minutes}.$$

It takes 0.45 s to charge the capacitor to its maximum value, but it takes 25.6 minutes to discharge it to 0.100 mC. This principle shows how a capacitor acts like a battery. When connected to loop 2, the resistor dissipates power which can light a lightbulb or heat a room. When switch $S1$ is opened, but switch $S2$ is not closed, the charge in the capacitor remains there because there is no closed circuit with a resistor to dissipate energy.

EXAMPLE 20.3

Consider the circuit shown in figure **EG-20.3**. The battery has an emf of 12.0 V. The resistors have resistance $R_1 = 250\,\Omega$ and $R_2 = 350\,\Omega$. The capacitor has capacitance $C = 150$ mF and is initially uncharged. Calculate the time constant of the circuit. (Hint: Use Kirchoff's rules).

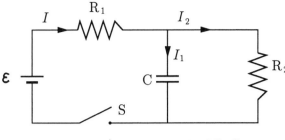

Figure: EG-20.3

SOLUTION

[Given: $\varepsilon = 12.0\,\text{V}$, $R_1 = 250\,\Omega$, $R_2 = 350\,\Omega$, and $C = 150\,\text{mF}$. Find: τ.]

Using Kirchoff's Loop rule for the right loop gives:

$$\frac{Q}{C} = I_2 R_2 \qquad (\text{and} \qquad Q_1 = \int I_1 \, dt)$$

And for the large loop, $\varepsilon = I R_1 + I_2 R_2$. Kirchoff's Node rule, at the top, says $I = I_1 + I_2$. So, using the first equation:

$$\int I_1 \, dt = C I_2 R_2$$

or

$$I_1 = C R_2 \frac{dI_2}{dt}$$

Substituting for I_1 in the node equation gives:

$$I = C R_2 \frac{dI_2}{dt} + I_2$$

Now, we need to guess (an educated guess) the form of I_2. We know at $t = 0$, $I_2 = 0$ and at $t = \infty$, $I_2 = I_{2\,max}$, so $I_2 = I_{2\,max}(1 - e^{-t/\tau})$ where τ is the time constant for the circuit.

173

Now we need to determine $I_{2\,max}$ and τ. At $t = 0$, $I = CR_2 \frac{dI_2}{dt}$ and $\frac{dI_2}{dt} = \frac{I_{2\,max}}{\tau} e^{-t/\tau}$ so $I = CR_2 \frac{I_{2\,max}}{\tau}$. Also, from the large loop equation: $\varepsilon = IR_1$ so $\frac{\varepsilon}{R_1} = CR_2 \frac{I_{2\,max}}{\tau}$.

And at $t = \infty$,

$$I = I_2. \quad \left(\text{Since, } \frac{dI_2}{dt} = \frac{I_{2\,max}}{\tau} e^{-t/\tau} = 0 \right)$$

so $I = I_{2\,max}(1 - e^{-t/\tau}) = I_{2\,max}$. Also, since $\varepsilon = IR_1 + I_2 R_2 = I(R_1 + R_2)$ so $I = \frac{\varepsilon}{R_1 + R_2}$ or $\frac{\varepsilon}{R_1 + R_2} = I_{2\,max}$.

Now substituting for $I_{2\,max}$ above gives

$$\frac{\varepsilon}{R_1} = CR_2 \frac{\left(\frac{\varepsilon}{R_1 + R_2} \right)}{\tau} \quad \text{or} \quad \tau = \frac{CR_1 R_2}{R_1 + R_2}$$

(Note: the units are correct $[\text{s}] = [\text{F}][\Omega]$) Now, substituting in the values gives:

$$\tau = \frac{(150 \times 10^{-3}\,\text{F})(250\,\Omega)(350\,\Omega)}{(250\,\Omega) + (350\,\Omega)} = 21.9\,\text{s}$$

So, the time constant for this circuit is $21.9\,\text{s}$.

Notice how we approached this problem. We used Kirchoff's rules to get us the initial equations. Then, we assumed a form for I_2, knowing that initially I_2 was zero. Then we applied the conditions $t = 0$ and $t = \infty$ to determine the constants. This procedure works for any complicated circuit.

EXERCISES

EXERCISE 20.1

Consider a capacitor and resistor in series with a switch as shown in figure **EX-20.1**.
The capacitor has a capacitance of 14.5 mF and initially holds a charge of 0.174 C.
The resistor has a resistance of 15.0 Ω. Calculate the charge on the capacitor, 2.00 s
after the switch is closed.

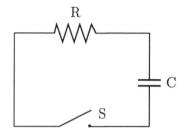

Figure: EX-20.1

EXERCISE 20.2

Consider a capacitor and a resistor in series with a switch, as shown in figure **EX-
20.2**. The capacitor has a capacitance of 19.2 mF and initially holds a charge of
0.275 C. The resistor has a resistance of 120 Ω. Calculate the current in the circuit
215 ms after the switch is closed.

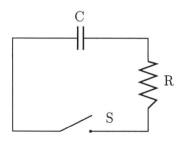

Figure: EX-20.2

EXERCISE 20.3

Consider a capacitor and a resistor in series with a switch as shown in figure **EX-20.3**. The capacitor has a capacitance of 20.5 mF and is initially charged with $Q = 0.275$ C. The resistor has a resistance of 250 Ω. Calculate the total heat generated by the resistor as the capacitor fully discharges.

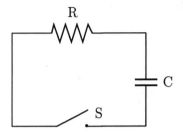

Figure: EX-20.3

EXERCISE 20.4

Consider a capacitor in series with a resistor, a battery, and a switch, as shown in figure **EX-20.4**. The capacitor has a capacitance of 23.2 mF and is initially uncharged. The resistor has a resistance of 735 Ω and the battery has an emf of 12.0 V. Calculate the current flowing in the circuit, 0.425 s after the switch is closed. [Start with the results from Example 20.1.]

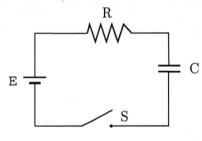

Figure: EX-20.4

EXERCISE 20.5

Consider a capacitor in series with a resistor, a battery, and a switch, as shown in figure **EX-20.5**. The capacitor has a capacitance of 41.5 mF and is initially uncharged. The battery has an emf of 9.00 V and the resistance of the resistor is 375 Ω. Calculate the total heat generated (work done) by the resistor as the capacitor completely charges.

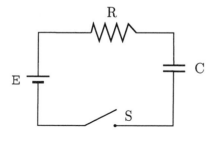

Figure: EX-20.5

EXERCISE 20.6

Consider the circuit shown in figure **EX-20.6**. If the capacitor initially holds a charge of 5.45 C, calculate the time required for the charge to decrease to 0.200 C. $R = 100 \, \Omega$ for each resistor. (Hint: Simplify the circuit before trying to solve it.)

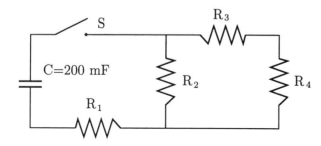

Figure: EX-20.6

EXERCISE 20.7

Consider the circuit shown in figure **EX-20.7**. Each capacitor is initially charged with $Q_i = 3.50 \, C$ and has a capacitance of $C = 300 \, mF$. The resistor has a resistance of 150 Ω. Calculate the time required until the combined charge on both capacitors decreases to 0.300 C.

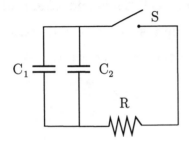

Figure: EX-20.7

EXERCISE 20.8

Consider the circuit shown in figure **EX-20.8**. Each capacitor is initially charged with $Q_i = 3.50\,\mathrm{C}$ and has a capacitance of $C = 300\,\mathrm{mF}$. The resistor has a resistance of $150\,\Omega$. Calculate the time required until the combined charge on both capacitors decreases to $0.300\,\mathrm{C}$.

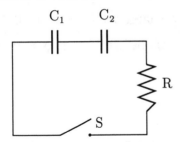

Figure: EX-20.8

EXERCISE 20.9

Consider a solar powered lamp. A possible model for the solar power system is shown in figure **EX-20.9**. The sun heats up the solar panels, and if the sun is directly shining on the solar panel, the emf is $220\,\mathrm{V}$. The $3.00\,\mathrm{F}$ capacitor is charged by the sun when switch $S1$ is closed. The resistance in the left loop is $120\,\Omega$. The switch $S1$ is closed for one hour during which time the sun shines directly on the solar panel. After one hour, the switch $S1$ is opened. When it gets dark, you want to use the light, so you close switch $S2$. Assuming you can see with the light until its power becomes less than 30 W, how long can you use the light if its resistance is 220 Ω.

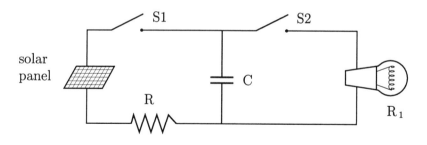

Figure: EX-20.9

EXERCISE   20.10

Consider the circuit shown in Figure **EX-20.10**. The battery has an emf of 15.0 V. The resistors have resistance $R_1 = 300\,\Omega$ and $R_2 = 425\,\Omega$. The capacitor has a capacitance of 215 mF and is initially uncharged. Find the current through resistor R_1 as a function of time.

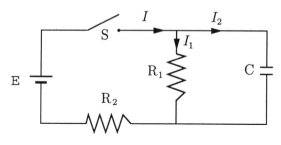

Figure: EX-20.10

EXAMPLE 21.1

A very long straight inhomogeneous rod of radius $R = 1\,\text{cm}$ carries a current density that increases linearly with a distance from the center of the rod according to: $J = ar$, where $a = 1.0\,\text{A/m}^3$. Determine the distribution of the magnetic field inside and outside the rod.

SOLUTION

[Given: inhomogeneous straight rod of radius $R = 1\,\text{cm}$, $J = ar$, where $a = 1\,\text{A/m}^3$. Find: **B**.]

The system has cylindrical symmetry so it is practical to use Ampere's law $\oint \mathbf{B}\cdot d\mathbf{l} = \mu_0 I = \mu_0 \int_{cs} \mathbf{J}\cdot d\mathbf{A}^*$ to calculate the magnetic field. Because of the cylindrical symmetry, the magnetic field lines are tangent to the circles concentric with the rod. Therefore, $\oint \mathbf{B}\cdot d\mathbf{l} = 2\pi r B$. Also, since the current density $\mathbf{J}$ points along the axis of the rod it is perpendicular to the cross-sectional area of the rod. This means that the surface element $d\mathbf{A}$ is collinear with the current density.

Let us now consider the case when the observation point is inside of the rod. The distance between the observation point and the axis of the rod is $r < R$ and only the current flowing through the portion of the cross-section of the rod contributes to the magnetic field:

$$\int_{cs} \mathbf{J}\cdot d\mathbf{A} = 2\pi a \int_0^r r^2\, dr = \frac{2\pi a r^3}{3}.$$

---

* we use *cs* to stand for "cross-csection".

Therefore, the magnitude of the magnetic field is

$$B = \frac{2\pi\mu_0 a r^3/3}{2\pi r} = \frac{a\mu_0}{3}r^2 = \frac{(1.0\,\text{A/m}^3)\mu_0}{3}r^2 = (4.2 \times 10^{-7}\,\text{T/m}^2)r^2 \ .$$

Next, consider the case of the observation point which is outside the current carrying rod. In this case the current density is integrated over the entire cross-section of the rod:

$$\int_{cs} \mathbf{J} \cdot d\mathbf{A} = 2\pi a \int_0^R r^2 \, dr = \frac{2\pi a R^3}{3} \ .$$

Therefore, the magnitude of the magnetic field is given by

$$B = \frac{2\pi a \mu_0 R^3/3}{2\pi r} = \frac{a R^3}{3r} = \frac{\mu_0 (1.0\,\text{A/m}^3)(0.01\,\text{m}^3)^3}{3r} = \frac{4.2 \times 10^{-13}\,\text{Tm}}{r} \ .$$

Note that the magnetic field is a continuous function and at $r = R$ takes the value $B(R) = \mu_0 a R^2/3 = 4.2 \times 10^{-11}\,\text{T}$.

EXAMPLE   21.2

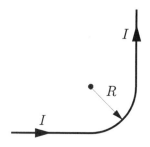

Figure: EG-21.2

Consider an infinitely long conducting wire that carries current $I = 20\,\text{A}$. The wire is bent as shown in Fig. EG-21.2. There are two segments which are at 90^0 angle relative to each. They are joined by a segment bent into a circular arc of radius $R = 10\,\text{cm}$. Find the magnetic field at the center of the circular arc.

[Given: infinitely long conducting wire carrying current $I = 20\,\text{A}$, bent as shown in figure EG-21.2. Find: $\mathbf{B}$ at the center of curvature of the circular segment.]

Using the right hand rule for the current flowing in the direction indicated in Fig. EG-21.2 it is apparent that the magnetic field points out of the plane of the wire. To calculate the magnetic field divide the conductor into three segments, two straight semiinfinite lines and an arc equal to a quarter of a circle. Using Biot-Savart law, $d\mathbf{B} = \mu_0 I (d\mathbf{l} \times \mathbf{r})/4\pi r^3$ it is easy to calculate the magnetic field of each current carrying segment. In Biot-Savart's law formula, the vector $\mathbf{r}$ points from the source point toward the observation point and the vector $d\mathbf{l}$ is along the current. It is practical to select the coordinate system such that the coordinate axis coincide with straight, semiinfinite wire segments as shown in Fig. EG-21.2a. In this coordinate system the center of the curvature of circular segment is at a location given by the vector $\mathbf{R} = -R\hat{x} + R\hat{y}$.

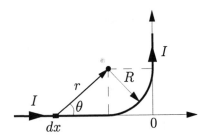

Figure: EG-21.2a

For the straight-line segment running along the negative portion of the x-axis, the line element is $d\mathbf{l} = dx\hat{x}$. The vector $\mathbf{r}$ pointing from the infinitesimal source segment at $x\hat{x}$ toward the observation point is given by $\mathbf{r} = \mathbf{R} - x\hat{x}$. The magnitude of the vector $\mathbf{r}$ is $r = \sqrt{(-R-x)^2 + R^2}$. The Biot-Savart's law reads,

$$d\mathbf{B} = \frac{\mu_0 I}{4\pi} \frac{dx\hat{x} \times (\mathbf{R} - x\hat{x})}{[(R+x)^2 + R^2]^{3/2}} \cdot$$

Because $\hat{x} \times \hat{x} = 0$ and $\hat{x} \times \hat{y} = \hat{z}$, the magnetic field at the observation point is given by

$$\mathbf{B}_1 = \hat{z}\frac{\mu_0 I R}{4\pi} \int\limits_{-\infty}^{R} \frac{dx}{((R+x)^2 + R^2)^{3/2}} \cdot$$

To evaluate the integral it is more convenient to use the variable θ indicated in Fig. EG-21-2a. Therefore, $x + R = -R\cot\theta$, (remember that x is negative for the segment running along the negative x-axis) and, $dx = Rd\theta/\sin^2\theta$. The integration bounds are simple in terms of the variable θ. The point at negative infinity, $x = -\infty$ corresponds to $\theta = 0$, and the point where the horizontal segment connects with the circular arc, $x = -R$, so, $\theta = 90^0 = \pi/2$. Also, $r = \sqrt{R^2\cot^2\theta + R^2} = R/\sin\theta$. Therefore, the integral reads

$$\mathbf{B}_1 = \hat{z}\frac{\mu_0 I R}{4\pi} \int\limits_{0}^{\pi/2} \frac{\sin^3\theta}{R^3} \frac{R}{\sin^2\theta} d\theta = \hat{z}\frac{\mu_0 I}{4\pi R} \int\limits_{0}^{\pi/2} \sin\theta \; d\theta$$

$$= \hat{z}\frac{\mu_0 I}{4\pi R} \cdot$$

The fact that $\int_0^{\pi/2} \sin\theta \; d\theta = 1$ was used.

Next, calculate the magnetic field of the vertical semiinfinite portion of the wire. It is not difficult, by following the logic used in calculating $\mathbf{B}_1$, to see that the contribution is the same. This time, $d\mathbf{l} = \hat{y}dy$ and an infinitesimal source segment is at a location $y\hat{y}$. Therefore, $\mathbf{r} = \mathbf{R} - y\hat{y}$. Hence, the magnetic field is given by an integral

$$\mathbf{B}_2 = \frac{\mu_0 I}{4\pi} \int\limits_{R}^{\infty} \frac{\hat{y}dy \times (\mathbf{R} - y\hat{y})}{[R^2 + (R-y)^2]^{3/2}} \cdot$$

Using, $\hat{y} \times \hat{y} = 0$ and $\hat{y} \times \hat{x} = -\hat{z}$ and changing to a new variable defined by

183

$y - R = R \cot \theta$ the integral evaluates to

$$\mathbf{B}_2 = \hat{z} \frac{\mu_0 I}{4\pi R} \; .$$

Finally, evaluate the contribution to the magnetic field of the current flowing through a circular arc section. To evaluate the magnetic field generated by this segment it is practical to consider a different coordinate system like the one depicted in Fig. EG-21.2b.

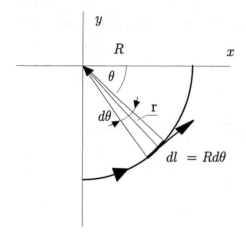

Figure: EG-21.2b

Then, $\mathbf{r} = -R(\cos\theta\hat{x} - \sin\theta\hat{y})$. The magnitude of the vector $\mathbf{r}$ is simply $r = R$. The line element vector $d\mathbf{l}$ is tangent to the arc of the circle and is given by $d\mathbf{l} = R(\cos\theta\hat{y} + \sin\theta\hat{x})d\theta$. Therefore, $d\mathbf{l} \times \mathbf{r} = R^2(-\cos^2\theta - \sin^2\theta)\hat{z}d\theta = -R^2\hat{z}d\theta$. Hence,

$$\mathbf{B}_3 = \frac{\mu_0 I}{4\pi} \int_{\pi/2}^{0} \frac{(-)R^2\hat{z}}{R^3} d\theta = \hat{z}\frac{\mu_0 I}{8R} \; .$$

Now that we have all of the contributions, the total magnetic field is given simply as a vector sum:

$$\mathbf{B} = \mathbf{B}_1 + \mathbf{B}_2 + \mathbf{B}_3$$
$$= \hat{z} \left(\frac{\mu_0 I}{4\pi R} + \frac{\mu_0 I}{4\pi R} + \frac{\mu_0 I}{8R} \right)$$
$$= \hat{z} \frac{\mu_0 I}{2\pi R} (1 + \pi) .$$

Using numerical values of the parameters, the magnitude of the magnetic field at the chosen point is $B = 714 \times 10^{-7} \, \text{T}$.

EXAMPLE 21.3

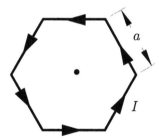

Figure: EG-21.3

Find the magnetic field of a wire bent into a form of a regular hexagon in the center of the hexagon. Assume that the current flowing through the current is $I = 10 \, \text{A}$ and that the hexagon has side of length $a = 20 \, \text{cm}$. See Fig. EG-21.3

SOLUTION

[Given: current carrying wire with $I = 10 \, \text{A}$ is bent into a shape of a regular hexagon of side $a = 20 \, \text{cm}$. Find: $\mathbf{B}$ in the center of the hexagon.]

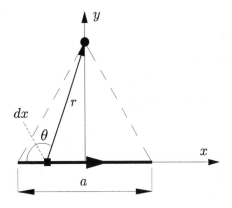

Figure: EG-21.3a

The geometry of the problem is given in Fig. EG-21.3. The hexagon can be thought of to consist of six straight segments of wire. Hence, to find the magnetic field of the hexagon at its center it is sufficient to find the magnetic field of one wire segment representing one side of the hexagon. The geometry for finding the magnetic field of one wire segment is given in Fig. EG-21.3a.

Consider a straight wire segment of length a carrying current I. The observation point is at the apex of an equilateral triangle. The observation point is distance $h = \sqrt{a^2 - (a/2)^2} = a\sqrt{3}/2$ above the wire. For definiteness, let us assume that the current flows in the direction as indicated in Fig. EG-23.3a. Select the coordinate system such that the wire lies along the x-axis and that the origin of the coordinate system is in the middle of the wire. Using Biot-Savart's law, $d\mathbf{B} = \mu_0 I (d\mathbf{l} \times \mathbf{r})/4\pi r^3$ it is not difficult to calculate the magnetic field. The line element is $d\mathbf{l} = \hat{x} dx$. The observation point is at a location $a\sqrt{3}\hat{y}/2$. The vector $\mathbf{r}$ which points from the infinitesimal segment at a location $x\hat{x}$ toward the observation point is, $\mathbf{r} = -x\hat{x} + a\sqrt{3}\hat{y}/2$. The magnitude of the vector $\mathbf{r}$ is $r = \sqrt{x^2 + 3a^2/4}$. The magnetic filed is given as an integral over the length of the wire segment,

186

$$\mathbf{B} = \frac{\mu_0 I}{4\pi} \int\limits_{-a/2}^{a/2} \frac{\hat{x} \times (-x\hat{x} + a\sqrt{3}\hat{y}/2)}{(x^2 + 3a^2/4)^{3/2}} \, dx \; .$$

Using the fact that $\hat{x} \times \hat{x} = 0$ and $\hat{x} \times \hat{y} = \hat{z}$, the formula reduces to

$$\mathbf{B} = \frac{\mu_0 I}{4\pi} \frac{a\sqrt{3}}{2} \hat{z} \int\limits_{-a/2}^{a/2} \frac{dx}{(x^2 + 9a^2/4)^{3/2}} \; .$$

To evaluate the integral introduce an angle variable θ indicated in Fig. EG-21.3a by a transformation $x = a\frac{\sqrt{3}}{2} \cot \theta$. Then, $dx = -a\sqrt{3}d\theta/2\sin^2\theta$. Also, $r = a\sqrt{3}/2\sin\theta$. The integration bounds in terms of the θ variable are simple. At $x = -a/2$, $\theta = 120^0 = 2\pi/3$; at $x = a/2$, $\theta = 60^0 = \pi/3$. The expression for the magnetic field becomes:

$$\mathbf{B} = \frac{\mu_0 I}{4\pi} \frac{a\sqrt{3}}{2} \hat{z} \int\limits_{2\pi/3}^{\pi/3} (-) \frac{a\sqrt{3}}{2\sin^2\theta} \left(\frac{2\sin\theta}{a\sqrt{3}} \right)^3 d\theta.$$

The remaining integral is simple, $-\int_{2\pi/3}^{\pi/3} \sin\theta \, d\theta = \cos\frac{\pi}{3} - \cos\frac{2\pi}{3} = 1$. Therefore, the magnetic field evaluates to:

$$\mathbf{B} = \frac{\mu_0 I \sqrt{3}}{6\pi a} \hat{z} \; .$$

Notice that all of the reference to the orientation of the wire segment in space has disappeared from the formula. The result depends only on the current through the wire and the location of the observation point relative to the wire. Therefore, the same result holds for any of the segments of the wire that form the sides of a hexagon. The net magnetic filed at the center of the hexagon is

$$\mathbf{B}_{total} = 6\mathbf{B} = 6\frac{\mu_0 I \sqrt{3}}{6\pi a}\hat{z}$$

$$= \frac{\mu_0 I \sqrt{3}}{\pi a}\hat{z} \ .$$

Using the numerical values for the parameters in the problem, the magnitude of the magnetic field in the center of the hexagon is

$$B_{total} = 34.6 \times 10^{-6}\,\mathrm{T} \ .$$

EXERCISES

EXERCISE 21.1

Find the energy stored in the magnetic field inside of a conducting cable of length $l = 1\,\mathrm{m}$ that has the radius $R = 1\,\mathrm{mm}$ if the current $I = 10\,\mathrm{A}$ flows through the conductor. Assume that the current is uniformly distributed. Neglect any effects due to the finite size of the cable!

EXERCISE 21.2

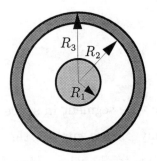

Figure: EX-21.2

188

A very long coaxial cable consists of two concentric conductors. The inner conductor is a solid cylinder of radius R_1 and the outer conductor is a hollow cylindrical shell of inner radius R_2 and outer radius R_3. The current flowing through the inner conductor has the same magnitude as the current flowing through the outer cable but they flow in opposite directions. Assume that the current density is uniform in each conductor. Find the magnetic field everywhere in space. See Fig. EX-21.2.

EXERCISE 21.3

Consider a straight piece of cable of length l carrying current I. Calculate the magnetic field generated by the cable at an arbitrary point in space outside the cable. You may assume that the cable is very thin so that its thickness can be neglected. [HINT: when setting up the problem you may find it helpful to use the coordinate axis such that the piece of a wire lies along one of the coordinate directions, lets say x-axis. Also, it is practical to chose the other direction, for example the y-axis such that it lies in the plane defined by the observation point and the conducting cable.

EXERCISE 21.4

Consider a current carrying cable of length L bent into a segment of an arc of a circle of radius R. Assume that the current flowing through the conductor is I. Find the magnetic field at the center of the circle.

EXERCISE 21.5

A current flows through an infinitely long straight conducting wire of circular cross-section of radius R. The current density is not uniform and changes according to $J = J_0 e^{-r/\delta}$, across the cross-section of the conducting wire. Find the magnetic field everywhere in space.

EXERCISE 21.6

Consider a particle of mass m and charge q moving in a constant magnetic field of magnitude B. Show that the trajectory of the particle is a circle. Find the radius of the circle. Let v_0 be the magnitude of the initial velocity of charged particle. Without loss of generality you may assume that the direction of the magnetic field is perpendicular to the direction of the initial velocity. [HINT: assume that the motion is circular and verify that the circular trajectory solves the equation of motion.]

EXERCISE 21.7

Consider a particle of mass m and charge q moving in a constant magnetic field of magnitude B. Let the initial velocity of the charged particle be $\mathbf{v}_0$ and without loss of generality you may assume that $\mathbf{v}_0$ is perpendicular to the direction of the magnetic field. Show that the magnetic force does no work on the particle, that is, the kinetic energy of the particle is unchanged.

EXERCISE 21.8

Consider a piece of a wire bent into a circle of radius R and carrying current I. The current loop is placed into a region of space with constant magnetic field pointing a direction perpendicular to the plane of the loop. What is the net force acting on the current loop?

EXERCISE 21.9

A current carrying wire in the form of a semicircle lies in a plane at right angles to the direction of a uniform magnetic filed. Show that the force on the wire is the same as that experienced by a straight wire lying along the diameter between the ends of the semicircle. See Fig. EX-21.9.

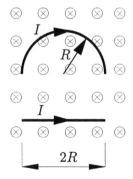

Figure: EX-21.9

EXAMPLE 22.1

Consider a U-shaped metal bar with a movable metal top cross-bar as shown in the figure. The setup which has a resistance of $20.0\,\Omega$ is standing so that the cross-bar can move vertically up and down and is placed in a location where a constant horizontal magnetic field $(B = 2.50\,\text{T})$ can act on the entire system. If the length of the cross-bar is $l = 0.500\,\text{m}$ and it has a mass of $2.00\,\text{kg}$, calculate the velocity of the cross-bar as a function of time after beginning at rest.

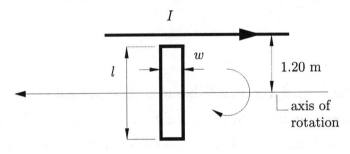

Figure: EX-22.1

SOLUTION

[Given: $R = 20.0\,\Omega$, $B = 2.50\,\text{T}$, $l = 0.500\,\text{m}$, $m = 2.00\,\text{kg}$, $v_0 = 0.00\,\text{m/s}$. Find: $v(t)$.]

First, we need to consider all the forces acting on the cross-bar. There is gravity and a magnetic force, which are opposing forces. So, $F_{total} = mg - IlB$ (downward). $I = \frac{\varepsilon}{R}$, where R is the resistance of the loop and ε is the emf. We can determine an expression for the emf, since

$$\varepsilon = -\frac{d}{dt}\Phi_m = -\frac{d}{dt}(B \cdot A) - B\frac{dA}{dt}.$$

The area is $l \cdot h$, where h is the height, so that

$$\frac{dA}{dt} = \frac{d}{dt}(l \cdot h) = l\frac{dh}{dt} = -lv,$$

(the height decreases as the bar falls). Now, substituting back in above, we get: $\varepsilon = -B(-lv) = Blv$ so $I = \frac{Blv}{R}$ and $F_{total} = mg - \frac{B^2l^2v}{R}$. We also know that $\Sigma F = ma = m\frac{dv}{dt}$, so

$$m\frac{dv}{dt} = mg - \frac{B^2l^2v}{R}$$

so $\frac{dv}{dt} = g - \frac{B^2l^2v}{mR}$ or $\frac{dv}{dt} + \left(\frac{B^2l^2}{mR}\right)v = g$. To solve this, we first solve the homogeneous equation:

$$\frac{dv}{dt} + \left(\frac{B^2l^2}{mR}\right)v = 0 \qquad \Rightarrow \int \frac{dv}{v} = \int -\left(\frac{B^2l^2}{mR}\right)dt$$

or $\ln v = -\left(\frac{B^2l^2}{mR}\right)t + C$ so that $v = Ae^{-\frac{B^2l^2}{mR}t}$ and A is a constant.

Now, back to the original differential equation, the solution should be $v = Ae^{-\frac{B^2l^2}{mR}t} + \frac{mRg}{B^2l^2}$. Let's check:

$$\frac{dv}{dt} + \left(\frac{B^2l^2}{mR}\right)v = g$$

$$-\frac{B^2l^2}{mg}Ae^{-\frac{B^2l^2}{mg}t} + \left(\frac{B^2l^2}{mR}\right)\left[Ae^{-\frac{B^2l^2}{mR}t} + \frac{mRg}{B^2l^2}\right] = g$$

Next, we determine A from the initial condition: at $t = 0$: $v = 0$, so, $0 = A + \frac{mRg}{B^2l^2}$ or $A = -\frac{mRg}{B^2l^2}$ so

$$v = \left(\frac{mRg}{B^2l^2}\right)\left[1 - e^{\frac{B^2l^2}{mR}t}\right]$$

Now, plugging in the values that were given:

$$v = \left[\frac{(2.0\,\text{kg})(20\,\Omega)(9.8\,\text{m/s}^2)}{(2.5\,\text{T})^2(0.50\,\text{m})^2}\right]\left[1 - e^{-\frac{(2.5T)^2(0.50\text{m})^2}{(2.0\text{kg})(20\Omega)}t}\right]$$

$$v = 251\,\text{m/s}\left(1 - e^{-t/(25.6\text{s})}\right)$$

Note: the units are a bit confusing. You might want to work them through to check them.

EXAMPLE 22.2

Consider a solenoid with $n = 210$ loops/meter, wound as shown in figure **EG-22.2**. A current source, of an R-C nature, produces a time varying current through the solenoid given by $I_1(t) = (6.5\,\text{A})e^{-t/(0.33\text{s})}$. The cross-sectional area of the solenoid is $0.40\,\text{m}^2$. A second loop, with only one turn, surrounds the solenoid near the center of the solenoid, and is connected to a resistor which has a resistance of $100\,\Omega$. This second loop has a cross-sectional area of $1.6\,\text{m}^2$. Find the current in the second loop as a function of time, $I_2(t)$.

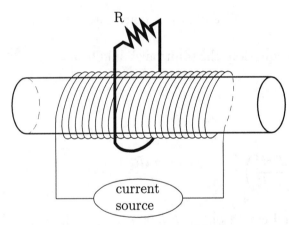

Figure: EG-22.2

SOLUTION

[Given: $n_1 = 210\frac{\text{turns}}{\text{meter}}$; $I_1(t) = I_0 e^{-t/\tau}$, where $I_0 = 6.5\,\text{A}$, and $\tau = 0.33\,\text{s}^{-1}$; $A_1 = 0.40\,\text{m}^2$, $R_2 = 100\,\Omega$, $A_2 = 1.6\,\text{m}^2$. Find: $I_2(t)$.]

Let's start by examining the second loopOC. The current in the loop depends on

194

its emf and its resistance:

$$I_2(t) = \frac{\varepsilon_2(t)}{R_2}$$

So, we need to determine $\varepsilon_2(t)$. We know that

$$\varepsilon_2(t) = -\frac{d\Phi_m}{dt}, \qquad \text{where} \qquad \Phi_m = \int \mathbf{B} \cdot d\mathbf{A} \ .$$

So, $\Phi_m = \int B \, dA$, since B only exists inside A_1. $\Phi_m = \int (\mu_0 n I_1) \, dA_1$, since $B = \mu_0 n I$ for a solenoid (see Chapter 21).

$$\Phi_m(t) = \mu_0 n I_1(t) A_1$$
$$= \mu_0 n A_1 (I_0 e^{-t/\tau})$$

Now,

$$\varepsilon(t) = -\frac{d}{dt}\Phi_m(t) = -\frac{d}{dt}\left[\mu_0 n A_1 I_0 e^{-t/\tau}\right]$$
$$= \frac{1}{\tau}\mu_0 n A_1 I_0 e^{-t/\tau}$$

So that $I_2(t) = \frac{\mu_0 n A_1 I_0}{\tau R_2} e^{-t/\tau}$ and, finally, substituting in the values:

$$I_2(t) = \frac{(4\pi \times 10^{-7}\text{Tm/A})(210 \text{ turns/m})(0.40\text{m}^2)(6.5\text{A})}{(0.33\text{s})(100\Omega)} e^{-\tau/0.33\text{s}}$$
$$= (2.08 \times 10^{-5}A)e^{-t/0.33\text{s}}$$
$$I_2(t) = (21\,\mu A)e^{-t/0.33\text{s}}$$

Notice that when we calculated the induced voltage $\varepsilon_2(t)$ that it only involved the area of loop 2 that contained a magnetic field. In other words, it only involved A_1, since the magnetic field is confined to be within the solenoid.

EXAMPLE 22.3

Consider the circuit shown below. The battery has an emf of 9.0 V, the resistors have resistance $R_1 = 300\,\Omega$, $R_2 = 150\,\Omega$, and the inductor has the inductance $L = 200\,\text{mH}$. Find the time constant of the circuit.

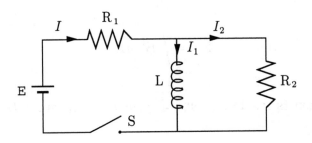

Figure: EG-22.3

SOLUTION

[Given: $\varepsilon = 9.0\,\text{V}$, $R_1 = 300\,\Omega$, $R_2 = 150\,\Omega$, $L = 200\,\text{mH}$. Find: τ.]

Start with Kirchoff's rules:

$$\text{The right loop:} \quad L\frac{dI_1}{dt} = I_2 R_2$$
$$\text{The large loop:} \quad \varepsilon = I R_1 + I_2 R_2$$
$$\text{The node rule:} \quad I = I_1 + I_2$$

Now, we need to pick a functional form for I_1: $I_1 = I_{1\,max}(1 - e^{-t/\tau})$ because at $t = 0$, $I_1 = 0$ and at $t = \infty$, $I_1 = I_{1\,max}$. So now, we know $\frac{dI_1}{dt} = \frac{I_{1\,max}}{\tau}e^{-t/\tau}$, so the first equation becomes: $L\frac{I_{1\,max}}{\tau}e^{-t/\tau} = I_2 R_2$.

Now, at $t = 0$: $I_2 - \frac{L \cdot I_{1\,max}}{R_2 \tau}$; $I = I_2$ and $\varepsilon = I_2(R_1 + R_2)$ so $\frac{\varepsilon}{R_1 + R_2} = \frac{L \cdot I_{1\,max}}{R_2 \cdot \tau}$. Then, at $t = \infty$: $I_2 = 0$, $I_1 = I_{1\,max}$, so $I = \frac{\varepsilon}{R_1} = I_1 = I_{1\,max} \Rightarrow I_{1\,max} = \frac{\varepsilon}{R_1}$. So, substituting in for $I_{1\,max}$ from the $t = 0$ information:

$$\frac{\varepsilon}{R_1 + R_2} = \frac{L\left(\frac{\varepsilon}{R_1}\right)}{R_2 \cdot \tau} \quad \text{or} \quad \tau = \frac{L(R_1 + R_2)}{R_1 R_2}$$

196

The units of τ are $[s] = \frac{[H]}{[\Omega]}$ which is correct.

Plugging in the numbers gives:

$$\tau = \frac{(200 \times 10^{-3}\text{H})(300\Omega)(150\Omega)}{300\Omega + 150\Omega} = 2.00 \times 10^{-3}\,\text{s}.$$

The time constant is, therefore, $2.00\,\text{ms}$.

EXERCISES

EXERCISE 22.1

Consider a long, straight current-carrying wire, as shown in the figure, with $I = 2.00\,\text{A}$ to the right. Below the wire is a long, very thin wire loop rotating into the page at the bottom of the loop. The loop has dimensions of $l = 1.50\,\text{m}$ and $w = 0.0200\,\text{m}$. The axis of rotation is $1.20\,\text{m}$ from the current-carrying wire. If the loop starts as shown and rotates at angular velocity of $60.0\,\text{rad/s}$, find the magnetic flux as function of time. (Assume the loop is narrow enough that B is constant along its width.)

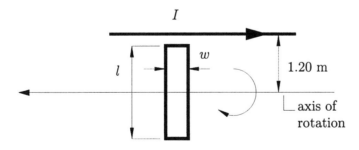

Figure: EX-22.1

EXERCISE 22.2

Consider a circular loop of wire, with area $0.350\,\mathrm{m}^2$, rotating in a constant magnetic field, $B = 1.31\,\mathrm{T}$, at a frequency of $3.00\,\mathrm{rotations/second}$. Calculate the induced emf as a function of time, if the loop is initially perpendicular to the field.

EXERCISE 22.3

Consider a loop of wire that has a bend in the middle as shown in figure **EX-22.3**. The area of each half of the loop is $0.500\,\mathrm{m}^2$. This loop starts as shown, with the lower part along the x-axis. A magnetic field points in the positive x-direction and remains constant at $B = 0.950\,\mathrm{T}$. If the wire loop rotates about the axis of the bend, with a constant angular velocity of $w = 1.00\,\mathrm{rotations/s}$, calculate the emf of the loop as a function of time.

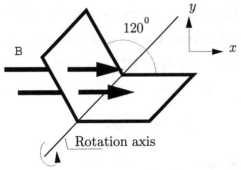

Figure: EX-22.3

EXERCISE 22.4

Consider a region in space where the magnetic field strength varies in the x-direction, so $B(x) = (0.300\,\mathrm{T/m})(6.00\,\mathrm{m} - x)$, as shown in figure **EX-22.4**. A square loop of wire with area $A = 0.0400\,\mathrm{m}^2$ starts with its left edge at $x = 0.00\,\mathrm{m}$ and travels to the right at a velocity of $0.100\,\mathrm{m/s}$. Calculate the emf of the loop as a function of time.

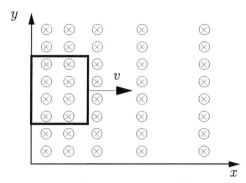

Figure: EX-22.4

EXERCISE 22.5

Consider a solenoid, with length $0.02\,\text{m}$ and 7 turns, wound as shown in figure **EX-22.5**. A current source produces a current through the solenoid that varies as: $I_1(t) = 6.5\,\text{A} + (2.1\,\text{A/s})t - (0.15\,\text{A/s}^2)t^2$. The cross-sectional area of the solenoid is $0.35\,\text{m}^2$. A second loop with only one turn, surrounds the solenoid near the center of the solenoid and is connected to a $250\,\Omega$ resistor. This second loop has an area of $1.35\,\text{m}^2$. Find the power output at the resistor as a function of time, $P_2(t)$.

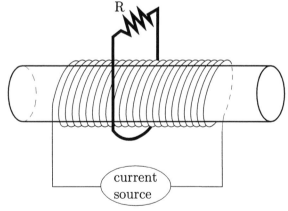

Figure: EX-22.5

EXERCISE 22.6

Consider the triangular metal wire, lying horizontally, shown in figure **EX-22.6**. A cross-bar sits on top the wire so that it makes electrical contact. The whole setup is immersed in a vertical magnetic field of strength 1.25 T. The cross-bar is pushed so that it moves toward the wide part of the triangle at a constant velocity of 0.10 m/s. Find the emf of this odd-shaped loop as a function of time, if the bar is initially 2.0 m away from the base of the triangle.

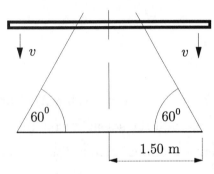

Figure: EX-22.6

EXERCISE 22.7

Consider a region where the magnetic field strength varies as $B(t) = (1.5\,\text{T})(1 - \frac{t}{20\text{s}})$, and a loop with $w = 0.20$ m and $l = 0.30$ m that is at the edge of this field at $t = 0$. If the loop moves to the left at a constant velocity of 2.0 m/s, calculate the emf of the loop during the first 0.10 s.

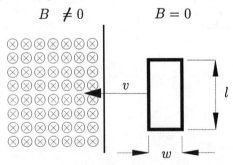

Figure: EX-22.7

EXERCISE 22.8

A tactical fighter plane is flying in a region where the earth's magnetic field is 5.00×10^{-5} T and points straight down. The plane flies with velocity of $v(t) = 200 \, \text{m/s} + (30.0 \, \text{m/s}^2)t$ and has a wing span of 20.0 m. The wing tips of this plane are vertical metal plates with a capacitance of 400 mF. Calculate the current flow from the negatively charged wing tip to the positively charged wing tip as a function of time.

EXERCISE 22.9

Consider an inductor connected in series with a resistor, a battery, and a switch, as shown in figure **EX-22.9**. The battery has an emf of 12.0 V, the resistor has a resistance of $1.50 \, \Omega$, and the inductor has an inductance of 70.0 mH. Calculate the current in the loop 30.0 ms after the switch is closed.

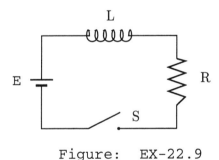

Figure: EX-22.9

EXERCISE 22.10

Consider the circuit shown in figure **EX-22.10**. The battery has an emf of $\varepsilon = 12.0$ V, the resistors have resistances of $R_1 = 15.0 \, \Omega$ and $R_2 = 25.0 \, \Omega$, and the inductor has an inductance of $L = 125$ mH. Calculate the power dissipated from resistor R_1 as a function of time.

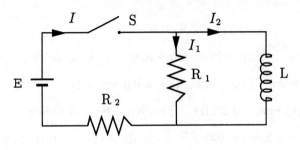

Figure: EX-22.10

Chapter 23   AC and Electronics

EXAMPLE   23.1

Consider an LC circuit. Assume that at time $t = 0$ the charge on the capacitor is Q_0 and that the current is zero. Write down the circuit equation and show that the current through the circuit and the charge on the capacitor oscillate in time. If $L = 1\,\text{mH}$ and $C = 10\,\mu\text{F}$, what is the frequency of oscillation? See Fig. EG-23.1.

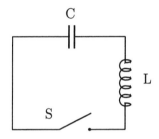

Figure: EG-23.1

SOLUTION

[Given: an LC circuit with $L = 1\,\text{mH}$ and $C = 10\,\mu\text{F}$. Show: Q and I oscillate in time; find the frequency ω of oscillations.]

The circuit is given in Fig. EG-23.1. To write down the circuit equation use Kirchoff's laws. The voltage change across the capacitor is $-Q/C$ and the voltage change across the inductor is $-L\,dI/dt$. The net voltage change in the circuit is zero for the closed loop. Hence,

$$-L\frac{dI}{dt} - \frac{Q}{C} = 0 \ .$$

Using the fact that the current is given as a time derivative of the charge, $I = dQ/dt$, the equation is transformed to:

$$\frac{dI}{dt} = \frac{d^2Q}{dt^2} = -\frac{1}{LC}Q.$$

This is the harmonic oscillator type equation with oscillation frequency $\omega = 1/\sqrt{LC}$. Numerically, $\omega = 1/\sqrt{(1 \times 10^{-3}\,\text{H})(10 \times 10^{-6}\,\text{F})} = 10^4\,\text{rad/s}$. To solve the equation try a solution of the form $Q = A\cos(\omega t + \delta)$. The current in the circuit is given by $I = dQ/dt = -\omega A \sin(\omega t + \delta)$. From the initial-conditions, $Q_0 = A\cos\delta$ and $0 = -A\omega\sin\delta$. Hence, $\delta = 0$ and $A = Q_0$. Therefore,

$$Q = Q_0 \cos\omega t\,, \qquad I = -I_0 \sin\omega t = I_0 \cos(\omega t + \pi.)\,,$$

where $I_0 = \omega Q_0$. Writing the current using the cosine function helps us see that the charge and the current are out of phase by 180 degrees!

EXAMPLE 23.2

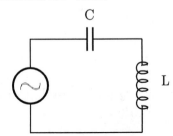

Figure: EG-23.2

Consider a driven LC circuit consisting of an inductor coil of inductance, $L = 1\,\text{mH}$, and a capacitor of capacitance, $C = 10\,\mu\text{F}$, connected to a source of alternating voltage, $\mathcal{E} = V_0 \cos\omega t$, with , $V_0 = 1\,\text{mV}$, and variable frequency ω. (a) What is the equation of the circuit? (b) What is the resonance frequency of the circuit. (c) Find the current in the circuit, and the charge on the plates of the capacitor as a function of time. (d) Describe the relationship of the voltage across the inductor and the voltage of the source and also the relationship between the voltage across the capacitor and the voltage of the source. See Fig. EG-23.2.

SOLUTION

[Given: driven LC circuit, $L = 1\,\text{mH}$, and $C = 10\,\mu\text{F}$, and a driving electromotive force $\mathcal{E} = V_0 \cos \omega t$, where, $V_0 = 1\,\text{mV}$, and variable driving frequency ω. Find: (a) equation of the circuit, (b) f_r, (c) $I(t)$ and $q(t)$, (d) describe the relation between $\mathcal{E}$ and V_L and $\mathcal{E}$ and V_C.]

(a) The circuit is depicted in Fig. EG-23.2. The circuit equation is obtained by applying Kirchoff's law's. The voltage drop across the capacitor is $V_c = -Q/C$. The voltage drop across the inductor coil is $V_L = -L dI/dt$. For a closed loop in the circuit, the net potential change zero. Hence, the circuit equation is

$$\mathcal{E} + V_C + V_L = 0 = \mathcal{E} - \frac{Q}{C} - L\frac{dI}{dt} \ .$$

Using the relation between the charge and the current, $I = dQ/dt$, yields an equation for the charge,

$$L\frac{d^2Q}{dt^2} + \frac{1}{C}Q = V_0 \cos \omega t \ .$$

(b) The circuit equation is an equation of a driven harmonic oscillator. Dividing the equation by L we read off that the natural frequency of the system is $\omega_0 = 1/\sqrt{LC} = 1 \times 10^4\,\text{rad/s}$. This means that the resonant frequency of the circuit is $f_r = \omega_0/2\pi = 1.6 \times 10^3\,\text{Hz}$.

(c) To find the solution of the circuit equation note that the equation is of the so called inhomogeneous type, i.e. there is a term in the equation (the driving term) which does not depend in the solution $Q(t)$. It is convenient to write the equation in the form

$$\frac{d^2Q}{dt^2} + \omega_0^2 Q = \frac{V_0}{L} \cos \omega t \ .$$

The general solution of the equation is a sum of the solution of the homogeneous equation (the equation without the source term) and a term which oscillates at a driving frequency. Write:

$$Q(t) = A \cos(\omega_0 t + \delta) + B \cos \omega t \ .$$

Two constants, A and δ, are determined by the initial conditions (which, by the way, were left unspecified in this problem). The second terms is a response to the driving term. The constant B is determined such the equation is satisfied. By inserting the assumed solution into the equation yields:

$$\frac{d^2Q}{dt^2} + \omega_0^2 Q = - A\omega_0^2 \cos(\omega_0 t + \delta) - B\omega^2 \cos \omega t + \omega_0^2 A \cos(\omega_0 t + \delta) + \omega_0^2 B \cos \omega t$$

$$= B(\omega_0^2 - \omega^2) \cos \omega t = \frac{V_0}{L} \cos \omega t \ .$$

Hence, the equation is satisfied if:

$$B = \frac{V_0/L}{\omega_0^2 - \omega^2} \ .$$

This result confirms that the frequency ω_0 is the resonant frequency. The current is easy to find by taking the derivative of the charge,

$$I = \frac{dQ}{dt} = -\omega_0 A \sin(\omega_0 t + \delta) - \frac{\omega V_0/L}{\omega_0^2 - \omega^2} \sin \omega t \ .$$

Notice that the first term in both, the charge and current functions, contains two unknown constants which must be determined from the initial conditions. The second term however, oscillates at the frequency of the driving force. In the event that the driving force frequency coincides with the natural or resonant frequency of the system, ω_0, the system is said to be in resonance and the amplitude of the charge and the current are divergent (infinite).

(d) Let us now calculate voltage drops across the inductor and the capacitor so that they can be compared with the driving electromotive force.

$$V_L = -L\frac{dI}{dt} = \omega_0^2 L A \cos(\omega_0 t + \delta) + \frac{V_0 \omega^2}{\omega_0^2 - \omega^2} \cos \omega t \, ,$$

$$V_C = -\frac{Q}{C} = -\frac{A}{C} \cos(\omega_0 t + \delta) - \frac{V_0 / \omega_0^2}{\omega_0^2 - \omega^2} \cos \omega t \, .$$

For comparison purposes, let also write the driving emf, $\mathcal{E} = V_0 \cos \omega t$. As long as the term that depends on the initial conditions is present, all voltages are completely out of phase with each other. It is more interesting to consider the situation when the homogeneous solution is not present which can be achieved by selecting the initial conditions such that $A = 0$.* Then:

$$V_L^{ss} = \frac{V_0 \omega^2}{\omega_0^2 - \omega^2} \cos \omega t \, ,$$

$$V_C^{ss} = -\frac{V_0 / \omega_0^2}{\omega_0^2 - \omega^2} \cos \omega t \, .$$

Therefore, the voltage across the inductor is in phase with the voltage of the electromotive force. However, because of, $-\cos \omega t = \cos(\omega t + \pi)$, the voltage across the capacitor is out of phase with the voltage of the driving source. When the voltage of the driving source and the voltage across the inductor are maximal, the voltage across the capacitor is minimal and *vice versa*.

---

* In a realistic case there will always be some resistance in the circuit. This resistance will make the homogeneous equation include a damping term. Therefore, a solution of the homogeneous equation will be damped after a sufficiently long time it will for practical purposes become negligible. The remaining solution is commonly called the steady state solution. Of course, do not forget that the presence of the resistance in the circuit changes the resonant frequency and the phase relations!

EXAMPLE 23.3

In a resonant circuit depicted in Fig. EG-23.3, a dissipative element, resistor R is connected in parallel with an inductor L and a capacitor C. Write down the equation for the circuit. Find the solution of the circuit equation. What is the resonance condition?

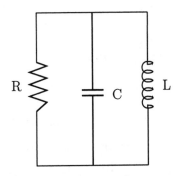

Figure: EG-23.3

SOLUTION

[Given: resonant circuit with a resistor of resistance R, an inductor of inductance L and a capacitor of capacitance C connected in parallel. Find: equation of the circuit; solution; resonance condition.]

Let us define the currents as in Fig. EG-23.3a. Let the voltage V be positive when the upper line is positive. Then, $Q = CV$, $I_1 = -dQ/dt = -CdV/dt$, $V = R(I_1 + I_2)$ and $V = -LdI_2/dt$. Using the equation for the voltage drop across the resistor the time derivative of the voltage reads, $dV/dt = R(dI_1/dt + dI_2/dt)$. Combining this relation with the expression for current I_1 and the expression for the time derivative of the current I_2 yields:

208

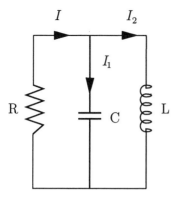

Figure: EG-23.3a

$$\frac{d^2V}{dt^2} + \frac{1}{RC}\frac{dV}{dt} + \frac{1}{LC}V = 0 \ .$$

Note that the equation can also be written in terms of the charge Q. To solve the equation we need to guess a trial solution. Also note that if the term with the first derivative of the potential was not present, the equation would be that of the harmonic oscillator system. The first derivative terms looks like a damping force term in the harmonic oscillator system. Therefore, as a trial solution, take:

$$V(t) = Ae^{-\alpha t}\cos\omega t \ ,$$

where α, ω and A are some constants. Next, calculate the first and the second derivative of the trial solution:

$$\frac{dV}{dt} = Ae^{-\alpha t}[-\alpha\cos\omega t - \omega\sin\omega t]$$

$$\frac{d^2V}{dt^2} = Ae^{-\alpha t}[(\alpha^2 - \omega^2)\cos\omega t + 2\alpha\omega\sin\omega t] \ ].$$

Substituting the derivatives into the circuit equation yields:

$$(\alpha^2 - \omega^2)\cos\omega t + 2\alpha\omega\sin\omega t - \frac{1}{RC}(\alpha\cos\omega t + \omega\sin\omega t)$$

$$+ \frac{1}{LC}\cos\omega t = 0 \ .$$

209

An overall exponential factor $e^{-\alpha t}$ was dropped. This equation will be satisfied if, and only if, the coefficients of $\sin \omega t$ and $\cos \omega t$ are both zero. In other words, require:

$$2\alpha\omega - \frac{\omega}{RC} = 0$$

and

$$\alpha^2 - \omega^2 - \frac{\alpha}{RC} + \frac{1}{LC} = 0 .$$

Solving for α yields, $\alpha = 1/2RC$. Hence, the circuit damping factor is $e^{-t/2RC}$. Next, solve for the frequency ω.

$$\omega^2 = \frac{1}{(2RC)^2} - \frac{1}{2(RC)^2} + \frac{1}{LC} = \frac{1}{LC} - \frac{1}{4R^2C^2} .$$

Hence, the circuit resonant frequency is

$$\omega_r = \sqrt{\frac{1}{LC} - \frac{1}{4R^2C^2}} .$$

The solution of the circuit is of the assumed form with α and ω determined above.

EXERCISES

EXERCISE 23.1

Consider a circuit in Fig. EX-23.1 where a capacitor of capacitance $C = 0.01\,\text{mF}$ and an inductor of inductance $L = 100\,\mu\text{H}$ and a resistor of resistance $R = 600\,\Omega$ are connected in series. Derive the equation of the circuit and show that the solution has the form $V(t) = Ae^{-\beta_1 t} + Be^{-\beta_2 t}$, which is appropriate for the overdamping. What is the ratio of constants A and B? Assume that at $t = 0$ the potential across the capacitor is $1\,\text{V}$ and that $I = 0$ at $t = 0$. Which term dominates?

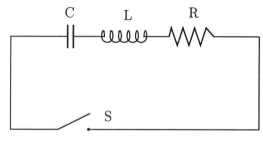

Figure: EX-23.1

EXERCISE 23.2

For a damped RLC circuit with a resistor R, an inductor L and a capacitor C connected in series find the expression for the total energy stored in the circuit (the sum of the energy stored in the capacitor and the energy stored in the inductor). Show that in the case of the critical damping, when $R = 2\sqrt{L/C}$, the energy is dissipated most quickly

EXERCISE 23.3

Consider a circuit consisting of a resistor R and an inductor L connected in series with an electromotive force $\mathcal{E} = V_0 \cos \omega t$. Find the steady state solution of the circuit.

EXERCISE 23.4

Consider a circuit consisting of a resistor R and a capacitor C connected in series with an electromotive force $\mathcal{E} = V_0 \cos \omega t$. Find the steady state solution of the circuit.

EXERCISE 23.5

In a circuit driven by an oscillating electromotive force at frequency ω the voltage across the resistor R is given by $V_0 \cos \omega t$. In general, the current will be out of phase with the voltage and has the form $I = I_0 \cos(\omega t + \phi)$. What is the average power delivered to the circuit.

EXERCISE 23.6

Consider a circuit depicted in Fig. EG-23.6. The circuit is driven by an electro-motive force $\mathcal{E} = V_0 \cos \omega t$. An inductor L and a resistor R are connected in series with each other and are connected in parallel with a capacitor C. What must be the relationship satisfied by the resistance R, the capacitance C, the inductance L and the frequency of the driving force so that the voltage of the driving electromotive force and the current in the circuit are in phase?

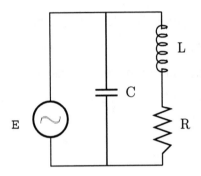

Figure: EG-23.3a

EXERCISE 23.7

In the US we commonly say that the outlet voltage is 120 V. This numerical value refers to the so called *root mean square* voltage. What is the amplitude of the AC voltage delivered to the power outlet assuming that it is a perfect sinusoidal voltage of frequency $f = 60$ Hz.

EXERCISE 23.8

In Europe and many other places in the world the common *rms* voltage is 220 V. If the voltage frequency is $f = 50$ Hz what is the voltage amplitude assuming that it is a perfect sinusoid.

EXAMPLE 24.1

Consider an electromagnetic plane wave which propagates in the $+x$-direction, as shown in figure **EG-24.1**. The magnetic field points along the $+z$-direction and has the form

$$B_z(x,t) = (2.00 \times 10^{-5}\,\text{T})\sin[(1.00 \times 10^7\,\text{m}^{-1})x - (3.00 \times 10^{15}\,\text{s}^{-1})t]$$

A loop, in the x-y plane, has a width $\Delta y = 1.00 \times 10^{-7}\,$m and a length of $\Delta x = 3.14 \times 10^{-7}\,$m. Calculate $\frac{d}{dt}\Phi_m$ through the loop as the plane wave propagates through it.

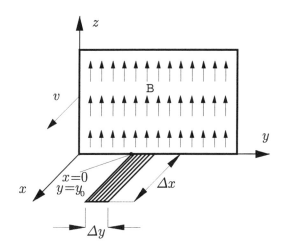

Figure: EG-24.1

SOLUTION

[Given: $B(x,t) = B_0\sin(kx - \omega t)$ where $B_0 = 2.00 \times 10^{-5}\,$T, $k = 1.00 \times 10^7\,\text{m}^{-1}$ and $\omega = 3.00 \times 10^{15}\,\text{s}^{-1}$; $\Delta y = 1.00 \times 10^{-7}\,$m, $\Delta x = 3.14 \times 10^{-7}\,$m. Find: $\frac{d\Phi_m}{dt}$.]

First, equation 22.1 tells us that: $\Phi_m = \int_{y_0}^{y_0 + \Delta y} \int_0^{\Delta x} \mathbf{B} \cdot d\mathbf{A}$ so

$$\Phi_m = \int_0^{\Delta x} B_0 \sin(kx - \omega t) \Delta y = -\Delta y B_0 \left(\frac{1}{k}\right) \cos(kx - \omega t) \Big|_0^{\Delta x}$$

To evaluate these limits, it is easier to split up the $\cos(A - B)$ term. Recall, $\cos(A - B) = \cos A \cos B + \sin A \sin B$. So,

$$\Phi_m = -\Delta y B_0 \left(\frac{1}{k}\right) [\cos kx \cos \omega t + \sin kx \sin \omega t] \Big|_0^{\Delta x}$$

$$= -\Delta y B_0 \left(\frac{1}{k}\right) [(\cos k\Delta x - \cos(0)) \cos \omega t + (\sin k\Delta x - \sin(0)) \sin \omega t]$$

Now, substituting for Δx notice that $\Delta x = \frac{\pi}{k}$, so $\cos k\Delta x = -1$ and $\sin k\Delta x = \sin \pi = 0$. So

$$\Phi_m = -\Delta y B_0 \left(\frac{1}{k}\right) [(-1 - (1)) \cos \omega t + 0]$$

$$= 2\Delta y B_0 \left(\frac{1}{k}\right) \cos \omega t$$

Now, we can differentiate with respect to time to get $\frac{d}{dt}\Phi_m$:

$$\frac{d}{dt}\Phi_m = \frac{d}{dt}\left[\frac{2\Delta y B_0}{k} \cos \omega t\right]$$

$$= -2\Delta y B_0 \left(\frac{\omega}{k}\right) \sin \omega t$$

Lastly, plugging in the numbers yields:

$$\frac{d}{dt}\Phi_m = -2(1 \times 10^{-7} \, \text{m})(2.0 \times 10^{-5} \, \text{T}) \left(\frac{3.00 \times 10^{15} \text{s}^{-1}}{1.00 \times 10^7 \text{m}^{-1}}\right) \sin(3.00 \times 10^{15} \text{s}^{-1}) t$$

$$\frac{d}{dt}\Phi_m = -(1.2 \times 10^{-3} \, \text{Tm}^2 \, /\text{s}) \sin(3.00 \times 10^{15} \, \text{s}^{-1}) t$$

EXAMPLE 24.2

Using the result of Example 24.1, and still considering the figure **EG-24.1**, calculate the induced electric field around the shaded loop as the plane wave propagates through it.

SOLUTION

[Given: $\frac{d}{dt}\Phi_m = -2\Delta y B_0(\frac{\omega}{k})\sin\omega t$, where $\Delta y = 1.00\times 10^{-7}$ m, $B_0 = 2.00\times 10^{-5}$ T, $\omega = 3.00\times 10^{15}$ s^{-1}, and $k = 1.00\times 10^7$ m^{-1}; $\Delta x = 3.14\times 10^{-7}$ m. Find: $E(t)$.]

We know from equation 22.6 that $\oint \mathbf{E}\cdot d\mathbf{l} = -\frac{d\Phi_m}{dt}$. Since we already calculated $\frac{d\Phi_m}{dt}$, all that remains is the line integral of $\mathbf{E}\cdot d\mathbf{l}$. Let's begin at the back corner of the loop, at $x = 0$ and $y = y_0$. Then, we will integrate counterclockwise around the closed loop (since the right hand rule would give B in the $+z$-direction if the loop is integrated counterclockwise).

$$\oint \mathbf{E}\cdot d\mathbf{l} = \int_0^{\Delta x} E_x(y = y_0)\,dx + \int_{y_0}^{y_0+\Delta y} E_y(x = \Delta x)\,dy$$

$$+ \int_{\Delta x}^{0} E_x(y = y_0 + \Delta y)\,dx + \int_{y_0+\Delta y}^{y_0} E_y(x = 0)\,dy \ .$$

To solve these integrals, let's re-examine our shaded loop. If $\mathbf{E}$ is a maximum at $x = 0$, then $\mathbf{E}$ is a minimum at $x = \Delta x$, since $\Delta x = \frac{\pi}{k}$, for any time t. In addition, $\mathbf{E}$ is not changing in the y-direction so $E(y = y_0) = E_x(y = y_0 + \Delta y)$. This tells us that being on the loop at $y = y_0$ is no different than being on the loop at $y = y_0 + \Delta y$, so $E_x(y = y_0) = E_x(y = y_0 + \Delta y)$. So, the two integrals over dx reduce to:

$$\int_0^{\Delta x} E_x(y = y_0)\, dx + \int_{\Delta x}^0 E_x(y = y_0 + \Delta y)\, dx = \int_0^{\Delta x} E_x\, dx + \int_{\Delta x}^0 E_x\, dx$$

$$= \int_0^{\Delta x} E_x\, dx - \int_0^{\Delta x} E_x\, dx = 0 \ .$$

Therefore, the integrals over dx cancel since they are integrated in opposite directions.

Now, $E_y(x = 0) \neq E_y(x = \Delta x)$, but E is constant over all y, so

$$\int_{y_0}^{y_0 + \delta y} E_y(x = \Delta x)\, dy = E_y(x = \Delta x) \int_{y_0}^{y_0 + \Delta y} dy = E_y(x = \Delta x)\, y \Big|_{y_0}^{y_0 + \Delta y}$$

$$= E_y(x = \Delta x)\, \Delta y$$

Similarly for the other integral:

$$\int_{y_0 + \Delta y}^{y_0} E_y(x = 0)\, dy = E_y(x = 0) \int_{y_0 + \Delta y}^{y_0} dy = E_y(x = 0)(-\Delta y).$$

Now, if $\mathbf{E}$ is a maximum at $x = 0$ then $\mathbf{E}$ is a minimum at $x = \Delta x$, or $E_y(x = 0) = -E_y(x = \Delta x)$, so, to put everything together:

$$\oint \mathbf{E} \cdot d\mathbf{l} = E_y(x = \Delta x)\Delta y - E_y(x = 0)\Delta y$$

$$= -E_y(x = 0)\Delta y - E_y(x = 0)\Delta y$$

$$= -2\Delta y E_y(x = 0)$$

And remember that $\oint \mathbf{E} \cdot d\mathbf{l} = -\frac{d}{dt}\Phi_B = -\left[-2\Delta y B_0 \left(\frac{\omega}{k}\right) \sin \omega t\right]$. So, $E_y(x = 0) = -B_0 \left(\frac{\omega}{k}\right) \sin \omega t = B_0 \left(\frac{\omega}{k}\right) \sin(-\omega t)$ and therefore, $E_y(x = \Delta x) = B_0 \left(\frac{\omega}{k}\right) \sin \omega t$. Since

E changes the direction by π radians over the length of the shaded loop, we suspect that **E** varies sinusoidally in the x-direction, so a possible form for E_y is:

$$E_y(x,t) = B_0 \left(\frac{\omega}{k}\right) \sin(kx - \omega t)$$

Notice in this problem that having the box a length of $x = \frac{\pi}{k}$ was very convenient. It is not necessary, however, to get this answer. In other words, this answer is independent of the size of the loop. If we had done the problem with a different sized loop, the math along the way would have been messier, but we would have ultimately obtained the same result.

So, plugging in the numbers:

$$E_y(x,t) = (2.0 \times 10^{-5}\,\text{T}) \left(\frac{3.00 \times 10^{15}\text{s}^{-1}}{1.00 \times 10^7 \text{m}^{-1}}\right) \sin[(1.00 \times 10^7\,\text{m}^{-1})x - (3.00 \times 10^{15}\,\text{s}^{-1})t]$$

$$E_y(x,t) = (6.00 \times 10^3\,\text{V/m}) \sin[(1.00 \times 10^7\,\text{m}^{-1})x - (3.00 \times 10^{15}\,\text{s}^{-1})t]$$

EXAMPLE 24.3

Now, with the result from Example 24.2, we would like to see if the electric field (induced by the original magnetic field), can then induce the magnetic field we began with in Example 24.1. (A self-sustaining electromagnetic wave is a little like the chicken and the egg. **B** induced **E**, which induced **B**, etc.) To check this, we begin with the same initial electromagnetic plane wave, and examine the effect of the electric field on a different loop. Our new loop lies in the x-z plane, as shown in figure **EG-24.3**, where $\Delta x = \frac{\pi}{k} = 3.14 \times 10^{-7}$ m and $\Delta z = 1.00 \times 10^{-7}$ m. Find the induced **B**-field from the original **E**-field calculated in Example 24.2.

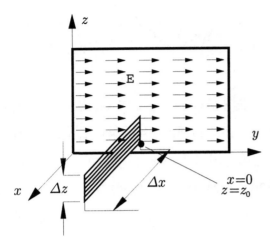

Figure: EG-24.3

SOLUTION

[Given: $\Delta x = \frac{\pi}{k} = 3.14 \times 10^{-7}$ m and $\Delta z = 1.00 \times 10^{-7}$ m, $E(x,t) = \frac{\omega}{k} B_0 \sin(kx - \omega t)$, where $\omega = 3.00 \times 10^{15}$ s^{-1}, $k = 1.00 \times 10^{-7}$ m^{-1}, and $B_0 = 2.00 \times 10^{-5}$ T. Find: $B(x,t)$.]

We begin by using equations 24.2 and 24.3 to get (in free space):

$\oint \mathbf{B} \cdot d\mathbf{l} = \mu_0 \, \varepsilon_0 \frac{d}{dt} \Phi_E$, where

$$\Phi_E = \int_{z_0}^{z_0+\Delta z} \int_0^{\Delta x} \mathbf{E} \cdot d\mathbf{A} = \Delta z \int_0^{\Delta x} E(x,t)\, dx = \Delta z \int_0^{\Delta x} \frac{\omega}{k} B_0 \sin(kx - \omega t)\, dx$$

$$\Phi_E = \Delta z \frac{\omega}{k^2} B_0 \left(\frac{-1}{k}\right) \cos(kx - \omega t)\Big|_0^{\Delta x}$$

$$= -\Delta z \frac{\omega}{k^2} B_0 \left[\cos kx \cos \omega t + \sin kx \sin(+\omega t)\right]\Big|_0^{\Delta x = \pi/k}$$

$$= -\Delta z \frac{\omega}{k^2} B_0 \left[(-1 - (+1)) \cos(+\omega t) + (0 - 0) \sin(+\omega t)\right]$$

$$= 2\Delta z \frac{\omega}{k^2} B_0 \cos(+\omega t) \qquad \text{and} \qquad \frac{d}{dt}\Phi_E = -2\Delta z \frac{\omega^2}{k^2} B_0 \sin(+\omega t)$$

Next, we need to evaluate the line integral $\oint \mathbf{B} \cdot d\mathbf{l}$. Again, the integral must be done counterclockwise around the loop because the electric field points in the $+y$-direction:

$$\oint \mathbf{B} \cdot d\mathbf{l} = \int_{z_0}^{z_0+\Delta z} B_z(x = 0)\, dz + \int_0^{\Delta x} B_x(z = z_0 + \Delta z)\, dx$$

$$+ \int_{z_0+\Delta z}^{z_0} B_z(x = \Delta x)\, dz + \int_{\Delta x}^0 B_x(z = z_0)\, dx$$

For similar reasons as discussed in Example 24.2:

$$\int_0^{\Delta x} B_x(z = z_0 + \Delta z)\, dx + \int_{\Delta x}^0 B_x(z = z_0)\, dx = 0$$

and $B_z(x = \Delta x) = -B_z(x = 0)$ which is constant in z, so

$$\oint \mathbf{B} \cdot d\mathbf{l} = \int_{z_0}^{z_0+\Delta z} B_z(x = 0)\, dz + \int_{z_0+\Delta z}^{z_0} B_z(x = \Delta x)\, dz$$

$$B_z(x = 0)\Delta z + B_z(x = \Delta x)(-\Delta z)$$

$$= 2\Delta z B_z(x = 0)$$

So, $2\Delta z B_z(x=0) = \mu_0\varepsilon_0\left[-2\Delta z\frac{\omega^2}{k^2}B_0\sin\omega t\right]$ or

$$B_z(x=0) = -(\mu_0\varepsilon_0)\left(\frac{\omega^2}{k^2}\right)B_0\sin\omega t$$

Lastly, $\mu_0\varepsilon_0 = \frac{1}{c^2}$, from equation 24.7, where c is the speed of propagation. Also, since c is the speed of propagation, $c = v = \frac{\omega}{k}$ so $\frac{\omega^2}{k^2} = c^2$. This leaves us with $B_z(x=0) = -\left(\frac{1}{c^2}\right)(c^2)B_0\sin\omega t$. Now, since $\sin A = -\sin(-A)$:

$$B_z(x=0) = B_0\sin(-\omega t)$$

And comparing to the original magnetic field:

$$B_z(x,t) = B_0\sin(kx-\omega t)$$

we see that at $x=0$ we get the same result as we expected.

EXERCISES

EXERCISE 24.1

In a circuit with a capacitor, current flows continuously through the wires, see figure **EX-24.1**. For a constant current I, show that the displacement current $\varepsilon_0\frac{d\Phi_E}{dt} = I$.

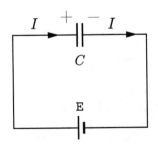

Figure: EX-24.1

220

EXERCISE   24.2

A circuit is made up of a capacitor, a resistor and a switch as shown in figure **EX-24.2**. The capacitor is initially charged with $8\,\mu C$ of charge, the capacitance is $7.5\,mF$ and the resistor has the resistance of $100\,\Omega$. Calculate the displacement current between the plates of the capacitor as a function of time.

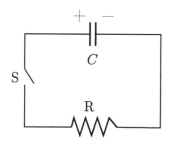

Figure: EX-24.2

EXERCISE   24.3

Now let's examine Maxwell's Displacement current in a capacitor with dielectric. In free space, the text explains that the dielectric current is $I_D = \varepsilon_0 \frac{d}{dt}\Phi_E = \varepsilon_0 A \frac{d}{dt}E$. In a dielectric, the equation needs to be modified. Show that the displacement current in a capacitor with a dielectric is $I_D = I = \varepsilon A \frac{d}{dt}E = \varepsilon \frac{d}{dt}\Phi_E$ by considering a dielectric which has $\varepsilon = 2.5\,\varepsilon_0$. This is the general form for the displacement current.

EXERCISE   24.4

Consider a circular, parallel plate capacitor, connected to a constant current source as shown in figure **EX-24.4**. The area of the plates is $A = 5.00 \times 10^{-3}\,m^2$ and the current flowing is $8.0\,mA$. The gap between the plates is $1.0 \times 10^{-3}\,m$. Show that the magnetic field generated between the plates is equal to the magnetic field generated around the wire.

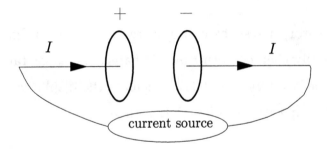

Figure: EX-24.4

EXERCISE   24.5

A circuit consists of a capacitor, a battery, a resistor and a switch as shown in figure **EX-24.5**. The capacitor has a capacitance of $10\,\mathrm{mF}$, the battery has $\varepsilon = 12\,\mathrm{V}$, and the resistor has the resistance of $250\,\Omega$. If the capacitor is initially uncharged, calculate the magnetic field induced between the plates.

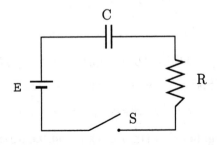

Figure: EX-24.5

EXERCISE   24.6

Consider an electric field that is confined to a circle of radius $10.0\,\mathrm{m}$. The electric field varies with time, but has the same value everywhere in the circle and points out of the page as shown in figure **EX-24.6**. The electric field goes as $\mathbf{E}(t) = (1500\,\mathrm{V/m})(\frac{t}{1\,\mu s})$. Find the induced magnetic field due to this configuration of the electric field. (Hint: the magnetic field is always perpendicular to the electric field, and in this case it is tangent to the circle.)

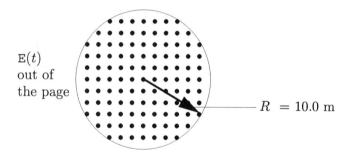

Figure: EX-24.6

EXERCISE   24.7

Calculate the irradiance of an electromagnetic wave, where

$$E(x,t) = (700\,\text{V/m})\sin[(1.00 \times 10^{-5}\,\text{m}^{-1})x - (3.00 \times 10^{13}\,\text{s}^{-1})t]\ .$$

EXERCISE   24.8

Given that a snake only perceives IR radiation between 700 nm and 800 nm, calcu-late the total irradiance perceived by a snake if it is 12.0 cm from a human. Assume that the irradiance perceived by the snake 12.0 cm from a human is:

$$I(\lambda) = 3.10 \times 10^{-10}\,\frac{\text{W}}{\text{m}^2}\,\frac{1}{(1000\,\text{nm} - \lambda)}\,d\lambda$$

EXERCISE   24.9

A laser beam has an amplitude for its electric field of:
$$E_0(x,y) = (750\tfrac{\text{V}}{\text{m}})e^{-(x^2+y^2)/4.00\times10^{-6}\text{m}^2}.$$

Calculate the total average power emitted by the laser beam.

EXERCISE 24.10

A spaceship powers its motion by the use of a radiation sail. A radiation sail is a large, thin metal sheet that reflects the electromagnetic radiation from a light source (such as the sun). This process transfers momentum to the ship (by reflection), giving it a velocity in the original direction of the electromagnetic wave. This spaceship and sail experiences an initial acceleration of $0.100\,\mathrm{m/s^2}$ when it is at rest at a distance of $0.500 \times 10^5\,\mathrm{km}$ from the sun. Find the velocity of the ship when it is $1.20 \times 10^5\,\mathrm{km}$ from the sun, neglecting for simplicity the sun's gravity. (Hint: the acceleration is not constant because the photon flux decreases with increasing distance from the sun.)

Chapter 28   SPECIAL RELATIVITY

EXAMPLE 28.1

Consider a proton, with mass 1.67×10^{-27} kg, charge 1.60×10^{-19} C, and initial velocity of 2.40×10^8 m/s in the $+x$-direction. This proton then enters a region in space with an electric field of 1000 V/m in the $-x$-direction. Calculate the initial acceleration of the proton. (Note: In chapter 24 we learned that charged particles emit gamma rays and other radiation to conserve energy when accelerating or decelerating. Ignore this effect in this problem.)

SOLUTION

[Given: $m = 1.67 \times 10^{-27}$ kg, $q = 1.60 \times 10^{-19}$ C, $\mathbf{v_0} = 2.4 \times 10^8$ m/s in $+x$-dir, and $\mathbf{E} = 1000$ V/m, in $-x$-dir. Find: a_0.]

We know that the force on the proton is $\mathbf{F} = q\mathbf{E}$ in $-x$-dir. So, $\mathbf{F} = q\mathbf{E} = \frac{d\mathbf{p}}{dt} = \frac{d}{dt}(\gamma m\mathbf{v}) = m[\gamma\frac{d\mathbf{v}}{dt} + \mathbf{v}\frac{d\gamma}{dt}]$. Now, recall that $\gamma = \left(1 - \frac{v^2}{c^2}\right)^{-1/2}$, so

$$\frac{d\gamma}{dt} = \frac{d}{dt}\left(1 - \frac{v^2}{c^2}\right)^{-1/2} = \left(1 - \frac{v^2}{c^2}\right)^{-3/2}\left(-\frac{1}{2}\right)\left(-\frac{2v\,dv}{c^2\,dt}\right) = \gamma^3\frac{v}{c^2}\frac{dv}{dt} = \gamma^3\frac{\beta}{c}\frac{d|v|}{dt},$$

where $\beta = v/c$. The force equation then becomes:

$$q\mathbf{E} = m\left[\gamma\frac{d\mathbf{v}}{dt} + \mathbf{v}\gamma^3\frac{\beta}{c}\frac{dv}{dt}\right].$$

Now, taking components of the vectors, since everything is in the x-direction:

$$qE_x = m\left[\gamma\frac{dv}{dt} + v\gamma^3\frac{\beta}{c}\frac{dv}{dt}\right].$$

225

Now, let's examine $\gamma + \gamma^3 \beta^2$:

$$\gamma + \gamma^3 \beta^2 = \frac{1}{(1 - \frac{v^2}{c^2})^{1/2}} + \frac{1}{(1 - \frac{v^2}{c^2})^{3/2}} \left(\frac{v}{c}\right)^2$$

$$= \frac{(1 - \frac{v^2}{c^2})}{(1 - \frac{v^2}{c^2})^{3/2}} + \frac{v^2/c^2}{(1 - \frac{v^2}{c^2})^{3/2}}$$

$$= \frac{1}{(1 - \frac{v^2}{c^2})^{3/2}} = \gamma^3 \ .$$

So,

$$\frac{qE_x}{m} = \gamma^3 \frac{dv}{dt}$$

or

$$\frac{dv}{dt} = \gamma^{-3} \frac{qE_x}{m} = \left(1 - \frac{v^2}{c^2}\right)^{3/2} \frac{qE_x}{m} \ .$$

So, plugging in the numbers gives:

$$\frac{d\mathbf{v}}{dt} = \mathbf{a} = \left[1 - \left(\frac{2.4 \times 10^8 \, \text{m/s}}{3.0 \times 10^8 \, \text{m/s}}\right)^2\right]^{3/2} \frac{(1.60 \times 10^{-19} \, \text{C})}{(1.67 \times 10^{-27} \, \text{kg})}$$

$$(-1000 \, \text{V/m}) \text{ in the } +x\text{-direction.}$$

$$\mathbf{a} = -2.07 \times 10^{10} \, \text{m/s}^2 \quad \text{–in the } +x\text{-dir.}$$

EXAMPLE 28.2

Consider the proton discussed in problem 28.1. Now that we know that the acceleration is described by $\frac{dv}{dt} = (1 - v^2/c^2)^{3/2} \frac{q}{m} E_x$ and $E_x = -1000 \, \text{V/m}$ in the $+x$-direction, calculate the speed of the proton as a function of time.

[Given: $\frac{dv}{dt} = \left(1 - \frac{v^2}{c^2}\right)^{3/2} \frac{q}{m} E_x$, $\mathbf{v}_0 = 2.4 \times 10^8 m/s$ in $+x$-dir, $m = 1.67 \times 10^{-27}$ kg, $q = 1.60 \times 10^{-19}$ C, $E_x = -1000$ V/m. Find: $v(t)$.]

We begin with $\frac{dv}{dt} = \left(1 - \frac{v^2}{c^2}\right)^{3/2} \frac{q}{m} E_x$ and integrate:

$$\int_{v_0}^{v_f} \frac{dv}{(1 - v^2/c^2)^{3/2}} = \int_0^t \frac{q}{m} E_x \, dt \quad \text{where} \quad \beta = \frac{v}{c}, \quad \text{so} \quad d\beta = \frac{1}{c} dv$$

$$\int_{v=v_0}^{v=v_f} \frac{c \cdot d\beta}{(1 - \beta^2)^{3/2}} = \frac{q}{m} E_x t \qquad \text{now let} \quad \beta = \cos\theta, \ d\beta = -\sin\theta \, d\theta$$

$$\int_{v=v_0}^{v=v_f} \frac{-\sin\theta \, d\theta}{(1 - \cos^2\theta)^{3/2}} = \frac{q}{mc} E_x t \qquad \text{but} \qquad (1 - \cos^2\theta) = \sin^2\theta$$

$$\int_{v=v_0}^{v=v_f} \frac{\sin\theta}{\sin^3\theta} d\theta = -\frac{q}{mc} E_x t \qquad \text{but} \qquad \sin^{-2}\theta = \csc^2\theta$$

$$\int_{v=v_0}^{v=v_f} \csc^2\theta d\theta = -\cot\theta \Big|_{v=v_0}^{v=v_f} = -\frac{q}{mc} E_x t \qquad \text{but} \qquad \cot\theta = \frac{\sin\theta}{\cos\theta} = \frac{\beta}{(1 - \beta^2)^{1/2}}$$

$$\frac{v_f/c}{(1 - (v_f/c)^2)^{1/2}} - \frac{v_0/c}{(1 - (v_0/c)^2)^{1/2}} = \frac{q}{mc} E_x t \ .$$

Now, to simplify this a little, let's rewrite and substitute in the numbers:

$$\frac{v_f}{[1-(v_f/c)^2]^{1/2}} = \frac{q}{m}E_x t + \frac{v_0}{[1-(v_0/c)^2]^{1/2}}$$

$$\frac{v_f}{[1-(v_f/c)^2]^{1/2}} = \frac{(1.60 \times 10^{-19}\text{C})}{(1.67 \times 10^{-27}\text{kg})}(-1000\text{V/m})\,t + \frac{(2.4 \times 10^8\text{m/s})}{\left[1 - \left(\frac{2.4 \times 10^8\text{m/s}}{3.0 \times 10^8\text{m/s}}\right)^2\right]^{1/2}}$$

$$\frac{v_f}{[1-(v_f/c)^2]^{1/2}} = -(9.58 \times 10^{10}\,\text{m/s}^2)t + 4.00 \times 10^8\,\text{m/s}.$$

Let $A = 9.58 \times 10^{10}\,\text{m/s}^2$ and $B = 4.00 \times 10^8\,\text{m/s}$.

$$v_f = (B - At)\left(1 - (v_f/c)^2\right)^{1/2}$$

$$v_f^2 = (B - At)^2(1 - (v_f/c)^2)$$

$$v_f^2 = (B - At)^2 - (B - At)^2\frac{v_f^2}{c^2}$$

$$\left[1 + \left(\frac{B - At}{c}\right)^2\right]v_f^2 = (B - At)^2$$

$$v_f = \sqrt{\frac{(B - At)^2 c^2}{c^2 + (B - At)^2}}$$

now, plugging in the values one last time:

$$v_f(t) = \frac{[4.00 \times 10^8\,\text{m/s} - (9.58 \times 10^{10}\,\text{m/s}^2)t](3.00 \times 10^8\,\text{m/s})}{\{(3.00 \times 10^8\,\text{m/s})^2 + [4.00 \times 10^8\,\text{m/s} - (9.58 \times 10^{10}\,\text{m/s}^2)t]^2\}^{1/2}}$$

$$v_f(t) = \frac{1.20 \times 10^{17}\,\text{m}^2/\text{s}^2 - (2.87 \times 10^{19}\,\text{m}^2/\text{s}^3)t}{\{9.00 \times 10^{16}\,\text{m}^2/\text{s}^2 + [4.00 \times 10^8\,\text{m/s} - (5.58 \times 10^{10}\,\text{m/s}^2)t]\}^{1/2}}.$$

EXERCISE 28.1

For relativistic corrections in problems, it is a useful mathematical observation that $f(x_0 + \Delta x) = f(x_0) + \frac{df}{dx}\big|_{x=x_0}\Delta x$. Use this observation to show that the non-relativistic kinetic energy, $\frac{1}{2}mv^2$, is the expression obtained from the relativistic kinetic energy, $\frac{mc^2}{\sqrt{1-(v/c)^2}} - mc^2$ if $\frac{v^2}{c^2}$ is small.

EXERCISE 28.2

To understand better the idea of time dilation, this problem investigates how time changes with velocity. For this problem, we want to calcualte the change in the unit of time, Δt, with respect to a small change in velocity. Calculate the change in the unit of time as the velocity changes from $0.9500c$ to $0.9501c$.

EXERCISE 28.3

To understand better the idea of length contraction, this problem investigates how the length of an object shrinks with velocity. For this problem, we want to calculate the change in length with respect to a small change in velocity. Calculate the change in the length of an object as the velocity changes from $0.9500c$ to $0.9501c$.

EXERCISE 28.4

Special relativity tells us that the force is $\frac{d}{dt}(\gamma mv)$ or $F = \gamma^3 ma$. Show that this is true by differentiating the kinetic energy with respect to x. $KE = mc^2(\gamma - 1)$.

Chapter 30   THE EVOLUTION OF QUANTUM THEORY

EXAMPLE   30.1

Consider an electron circling around a proton, as in Bohr's description of a hydrogen atom. Bohr suggested that the electron travels in a circular orbit, so that $F_c = \frac{mv^2}{r}$ and the potential energy of the electron is $PE = \frac{kq_eq_p}{r} = -\frac{kq^2}{r}$. Show that the kinetic energy is: $KE = -\frac{1}{2}PE$.

SOLUTION

[Given: $F = \frac{mv^2}{r}$, $PE = -\frac{kq^2}{r}$. Find: KE.]

We know that the kinetic energy is $KE = \frac{1}{2}mv^2$, so re-writing the KE in terms of F gives:

$$KE = \frac{1}{2}\frac{mv^2}{r}r = \frac{r}{2}F \ .$$

And $F = \frac{dPE}{dr}$, where $PE = -\frac{kq^2}{r}$, so

$$KE = \frac{r}{2}\frac{d}{dr}\left(\frac{-kq^2}{r}\right) = \frac{r}{2}\frac{kq^2}{r^2} = -\frac{1}{2}\left(-\frac{kq^2}{r}\right) = -\frac{1}{2}PE \ .$$

So the kinetic energy equals $KE = -\frac{1}{2}PE$.

EXERCISE   30.1

A cube of steel is 0.200 m on a side and has been heated to 400^0 C. Its total emissivity is 0.970 and its temperature decreases as $T(t) = (673\,\text{K})\left(1 - \frac{t}{65.0\text{s}}\right)$ for the first 10.0 seconds after it is removed from the furnace. Calculate the total energy emitted during these first 10.0 seconds.

In Compton scattering, we find that $\Delta\lambda = \frac{h}{m_e c}(1 - \cos\theta)$. Use calculus to find the angle that produces the maximum value for $\Delta\lambda$.

EXAMPLE   32.1

We know that $\frac{dN}{dt} = -\lambda N$, where λ is the decay constant. We also know that $R = \lambda N$, where R is the decay rate. Derive the differential equation for R, and using that equation, derive equation (32.11).

SOLUTION

[Given: $\frac{dN}{dt} = -\lambda N$ and $R = \lambda N$. Find: $\frac{dR}{dt}$ and $R(t)$.]

Start with the equation given: $\frac{dN}{dt} = -\lambda N$. Now, since $R = \lambda N$, substitute for N ($N = R/\lambda$).

$$\frac{d}{dt}\left(\frac{R}{\lambda}\right) = -\lambda\left(\frac{R}{\lambda}\right) .$$

But, λ is a constant, so $\frac{d}{dt}\left(\frac{R}{\lambda}\right) = \frac{1}{\lambda}\frac{dR}{dt}$, giving

$$\frac{1}{\lambda}\frac{dR}{dt} = -\lambda R\left(\frac{1}{\lambda}\right) \qquad \text{or} \qquad \frac{dR}{dt} = -\lambda R .$$

This is the differential equation for R. Now, to solve this equation, we must re-write it:

$$\int_{R_0}^{R} \frac{dR}{R} = \int_{0}^{t} -\lambda\, dt$$

$$\ln R\Big|_{R_0}^{R} = -\lambda t$$

$$\ln\left(\frac{R}{R_0}\right) = -\lambda t$$

Now, we exponentiate both sides of the equation:

$$R = R_0 e^{-\lambda t},$$

which is equation 32.11.

EXAMPLE 32.2

Consider a sample of cobalt-60, which initially has a decay rate of 2.00×10^{10} decays/day. The half-life of Co-60 is 5.24 yr. Calculate the number of Co-60 atoms remaining after 24.0 years.

SOLUTION

[Given: $R_0 = 2.00 \times 10^{10}$ decays/day, $t_{1/2} = 5.24$ yr. Find: N at $t = 24.0$ yr.]

First, we can determine the decay rate after 24.0 yr. So, $\frac{dR}{dt} = -\lambda R$ or $\int_{R_0}^{R} \frac{dR}{R} = \int_0^t -\lambda \, dt$ and $\lambda = \frac{0.693}{t_{1/2}}$. We have after integrating, $\ln R \big|_{R_0}^{R} = \ln \frac{R}{R_0} = -\lambda t$ so

$$R = R_0 e^{-\lambda t} = R_0 \, e^{-\left(\frac{0.693}{t_{1/2}}\right)t} \, .$$

Notice that $t_{1/2}$ is a constant and therefore does not need to be inside the integral. Now, plugging in values,

$$R = (2.00 \times 10^{10} \text{ decays/day}) \, e^{-\frac{0.693 \,(24.0 \text{yr})}{5.24 \text{ yr}}} = 8.36 \times 10^8 \text{decays/day}.$$

Next, since $R = \lambda N$, $N = \frac{R}{\lambda} = \frac{R t_{1/2}}{0.693}$

$$N = (8.36 \times 10^8 \text{ decays/day}) \frac{(5.24 \text{ yr})}{0.693} \left(\frac{365.25 \text{ days}}{1 \text{ yr}} \right)$$

$$N = 2.31 \times 10^{12} \text{ atoms} \qquad \text{remains after 24.0 years} \, .$$

EXERCISE 32.3

The shroud of Turin, which for decades was purported to be the burial shroud of Jesus Christ, was recently tested by carbon dating techniques to determine its age. The laboratories calculated that the shroud was approximately 1320 years old. Carbon dating determines the approximate age by measuring the ratio of ^{14}C to ^{12}C. If a living organism has a ratio of 1.3×10^{-12}, calculate the ratio of ^{14}C to ^{12}C that was measured by the laboratories. $(t_{1/2}(^{14}\text{C}) = 5730\,\text{y.})$

SOLUTION

[Given: $t = 1320\,\text{yr}$, ratio$_i = 1.3 \times 10^{-12}$, $t_{1/2} = 5730\,\text{yr}$. Find: ratio$_f$.]

Since the number of ^{12}C atoms does not change, the ratio is basically the number of ^{14}C atoms divided by a constant, so we are looking for the final number of ^{14}C atoms. We know $\frac{dN}{dt} = -\lambda N$ and $\lambda = \frac{0.693}{t_{1/2}}$, so

$$\int_{N_0}^{N_f} \frac{dN}{N} = -\int_0^t \lambda\, dt$$

$$\ln N \Big|_{N_0}^{N_f} = -\lambda t$$

$$\ln \left(\frac{N_f}{N_0} \right) = -\lambda t$$

$$\text{so} \quad N_f = N_0 e^{-\lambda t}$$

or

$$\frac{N_f^{14}\text{C}}{N^{12}\text{C}} = \frac{N_i^{14}\text{C}}{N^{12}\text{C}} e^{-\lambda t}$$

so ratio$_f$ = ratio$_i e^{-\lambda t}$ = ratio$_i\, e^{-\left(\frac{0.693}{t_{1/2}}\right)t}$. Lastly, let's plug in the numbers:

$$\text{ratio}_f = (1.3 \times 10^{-12}) e^{-\left(\frac{0.693}{5730\,\text{yr}}\right)(1320\,\text{yr})}$$

$$\text{ratio}_f = 1.11 \times 10^{-12}.$$

So the laboratories must be able to measure the ratio to at least two significant figures. The traditional way of determining this ratio would have required a *large* amount of the shroud to be destroyed, but that would not be allowed. In 1988, the process had been refined so that only a one-inch square was needed. That was acceptable to the bishop of Turin, so the test was then performed.

EXERCISES

EXERCISE 32.1

Consider a sample of fluorine-17, which has a half-life of 66 s. If we initially have 3.00×10^{30} atoms of F-17, what is the decay rate after one hour?

EXERCISE 32.2

Consider a sample of iodine-131, which has a half-life of 8.0 days. If the sample has an initial decay rate of 5.25×10^{10} decays/day, calculate the decay rate exactly one week later.

EXERCISE 32.3

A sample of radon-222, contains 3.00×10^8 atoms of radon today. The half life of Rn-222 is 3.82 days. How many atoms of radon were in the sample 30 days ago?

EXERCISE 32.4

You want to determine the half-life of a material, so you measure its activity at two different times. At $t = 0.0$ hr you measure a decay rate of 3.00×10^8 decays/s. At $t = 1.00$ hr, you measure a decay rate of 2.00×10^8 decay/s. What is the half-life of this material?

EXERCISE 32.5

You are given the task of determining the half-life of an unknown isotope. You are told that after exactly one hour, the number of atoms of this isotope decreased from 5.000×10^{15} atoms to 4.4963×10^{14} atoms. Use the expression $\frac{dN}{dt} = -\lambda N$

and calculus to determine the half-life of this radioisotope.

EXERCISE 32.6

Consider a sample of phosphorus-14, which has a half life of 14 days. You measure the decay rate to be 5.00×10^{15} decays/hr, 18 days after the sample was created. What was the initial decay rate of the sample?

EXERCISE 32.7

A sample of radioactive material initially contains 10.0 kg of uranium-235. Given that the atomic mass of U-235 is 235.043924 u and that its half life is 7.038×10^8 y, calculate the time required until there is 9.50 kg of U-235 in the sample. Use the expression $\frac{dN}{dt} = -\lambda N$ and calculus to solve the problem.

EXERCISE 32.8

When digging in your yard, you find a very old piece of wood. You bring it into your physics class to determine how old it is. The analysis tells you that the ratio of ^{14}C to ^{12}C in the wood is 1.0×10^{-12}. When the wood was *alive* as a tree, the ratio of ^{14}C to ^{12}C was 1.3×10^{-12}. If the half life of ^{14}C is 5730 y, how long has the wood been dead?

EXERCISE 32.9

In reality, the number of radioactive atom obeys the equation $\frac{dN}{dt} = -\lambda N$. What if, instead, the decay went as $\frac{dN}{dt} = -BNt$, where B is a constant? Calculate the functional form of $N(t)$, and calculate an expression for the half-life.

EXERCISE 32.10

In reality, the number of radioactive atoms obeys the equation $\frac{dN}{dt} = -\lambda N$. What if, instead, the decay went as $\frac{dN}{dt} = -AN^{1/2}$, where A is a constant? Calculate the functional form for the number of atoms as a function of time. What is the expression for the half life now?

Chapter 2   Kinematics: Speed and Velocity

Answer to Problem: 2.1

17.1 cm/s.

Answer to Problem: 2.2

(a) $v_{av} = 70\,\text{cm/s}$; (b) $v = 80\,\text{cm/s}$

Answer to Problem: 2.3

(a) $\mathbf{v}(0\,\text{s}) = 14.7\,\text{m/s-toward the stop sign}$; (b) $t = 6.9\,\text{s}$

Answer to Problem: 2.4

$v_{\text{max}} = 11.5\,\text{m/s}$

Answer to Problem: 2.5

(a) $v(t) = (6.0\,\text{m/min}^2)t - (1.5\,\text{m/min}^3)t^2 + 7.5\,\text{m/min}$; (b) $t = 5.0\,\text{min}$.

Answer to Problem: 2.6

$v = 5.0\,\text{m/s}$ at $r = 10\,\text{m}$

Answer to Problem: 2.7

$v(t) = (1.0\,\text{m/s}^2)t + 2.7\,\text{m/s}$

Answer to Problem: 2.8

(a) $t = 5.27\,\text{s}$ (b) $t = 6.69\,\text{s}$, $v = 162\,\text{km/hr}$

ANSWER TO PROBLEM: 2.9

$$v_{AB} = -9.2 \, \text{m/s} + (4.0 \, \text{m/s}^2)t$$

ANSWER TO PROBLEM: 2.10

The speed of the dropping water is 0.53 cm/s when the water is at a height of 2.0 cm.

Chapter 3   KINEMATICS: ACCELERATION

ANSWER TO PROBLEM: 3.1

(a) The motion is two-dimensional. Velocity components v_x and v_y are linear functions of time. Two constants, α and γ are measured in a unit of distance divided by a unit of time. They are the components of the initial velocity vector; $v_{0x} = \alpha$, $v_{0y} = \gamma$. Constants β and δ are measured in a unit of length divided by the square of a unit of time. They are the components of the acceleration vector; $a_x = \beta$, $a_y = \delta$.

$x(t) - x_0 = \alpha + \beta t^2/2$, $y - y_0 = \gamma + \delta t^2/2$ and x_0 and y_0 are the initial positions in x and y directions, respectively.

(c) $a_x = dv_x/dt = \beta$ and $a_y = dv_y/dt = \delta$.

(d) $v = \sqrt{v_x^2 + v_y^2} = \sqrt{(\alpha + \beta t)^2 + (\gamma + \delta t)^2}$, $\tan\theta_v = v_y/v_x = (\gamma + \delta t)/(\alpha + \beta t)$, where θ_v is the angle that the velocity vector makes with the x-axis;
$a = \sqrt{a_x^2 + a_y^2} = \sqrt{\beta^2 + \delta^2}$, $\tan\theta_a = a_y/a_x = \delta/\beta$, where θ_a is the angle that the acceleration vector makes with the x-axis.

ANSWER TO PROBLEM: 3.2

$v_x(t) = dx/dt = b_x + 2c_x t + 3d_x t^2$ and $v_y(t) = dy/dt = b_y + 2c_y t + 3d_y t^2$;
$v_x(0) = b_x$ and $v_y(0) = b_y$.

$a_x(t) = d^2x/dt^2 = 2c_x + 6d_x t$ and $a_y(t) = d^2y/dt^2 = 2c_y + 6d_y t$;
$a_x(0) = 2c_x$ and $a_y(0) = 2c_y$.

ANSWER TO PROBLEM: 3.3

(a) $v_{av} = \frac{\Delta x}{\Delta t} = \frac{1}{t}\int_0^t v(t)dt = v_0 t_f + at/2 = (v_0 + v_f)/2.$

(b) $v(t) = \begin{cases} v_0 + at & \text{for } 0 \leq t \leq t_f; \\ v_f = v_0 + at_f & \text{for } t_f < t \leq t_f + t_{add} . \end{cases}$

239

and $v_{av} = \frac{1}{t+t_a} \int_0^{t+t_a} v(t)dt = \frac{v_0 t_f + \frac{1}{2}at_f^2}{t_f + t_{add}} + \frac{v_f t_{add}}{t_f + t_{add}}$.

Results in parts (a) and (b) are clearly different.

ANSWER TO PROBLEM: 3.4

$x_m = v_0/\sqrt{\alpha} = 0.016\,\mathrm{m}$.

ANSWER TO PROBLEM: 3.5

$a = \alpha t = dv/dt$; $v = v_0 + \alpha t^2/2 = ds/dt$; $s = s_0 + v_0 t + \alpha t^3/6$.

ANSWER TO PROBLEM: 3.6

To understand answers better, make a plot of velocity as a function of time. $a = dv/dt = -(2\,\mathrm{m/s^3})t + 8\,\mathrm{m/s^2}$.

(a) The velocity increases from $t = 0\,\mathrm{s}$ to $t = 4\,\mathrm{s}$. The speed is increases in the same time interval. In addition, the speed also increases from $t = 8\,\mathrm{s}$ to $10\,\mathrm{s}$.

(b) The peak velocity occurs at time $t = 4\,\mathrm{s}$. The maximal velocity is, $v(4\,\mathrm{s}) = 16\,\mathrm{m/s}$.

(c) The speed peaks at $t = 10\,\mathrm{s}$. At $t = 10\,\mathrm{s}$ the velocity equals to $-20\,\mathrm{m/s}$ which means that the speed at that moment is $+20\,\mathrm{m/s}$.

ANSWER TO PROBLEM: 3.7

$v_x = dx/dt = 8\,\mathrm{m/s}$, $v_y = dy/dt = (40\,\mathrm{m/s^2})t$, $v_z = dz/dt = 0$;
$a_x = dv_x/dt = 0$, $v_y = dv_y/dt = (40\,\mathrm{m/s^2})$, $a_z = dv_z/dt = 0$; parabola

ANSWER TO PROBLEM: 3.8

$v_x = dx/dt = (180\,\mathrm{m/s})\cos(15\,\mathrm{s^{-1}})t$, $v_y = 25\,\mathrm{m/s}$;
$a_x = dv_x/dt = -(2700\,\mathrm{m/s^2})\sin(15\,\mathrm{s^{-1}})t$, $a_y = dv_y/dt = 0$.

ANSWER TO PROBLEM: 3.9

(a) The plot is given in figure EX-3.9.

(b) $v_{av} = \frac{1}{1\,\text{s}} \int_2^3 dt\ (10\,\text{m})e^{-0.2t} = -\frac{10\,\text{m}}{1\,\text{s}} \frac{1}{0.2\,\text{s}^{-1}} e^{-0.2t} \Big|_2^3 = 6.1\,\text{m/s}.$

(c) $a = \frac{d^2 x}{dt^2} = (0.4\,\text{m/s}^2)e^{-0.2t}.$

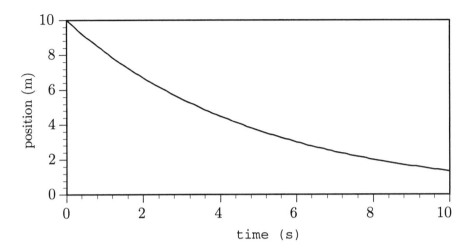

Figure: EX-3.9

ANSWER TO PROBLEM: 3.10

$v_x = dx/dt = (2\,\text{m/s})e^{(2\,\text{s}^{-1})t},\ v_z = dz/dt = 4.0\,\text{cm/s}^2;$
$a_x = dv_x/dt = (4\,\text{m/s}^2)e^{(2\,\text{s}^{-1})t},\ a_z = dv_z/dt = 0.$

Answer to Problem: 4.1

impulse= 72.8 kg m/s

Answer to Problem: 4.2

$F = (240\,\mathrm{N/s})(t - 1\,\mathrm{s})$. After 1 second you get your dog back in control.

Answer to Problem: 4.3

(a) The impulse delivered to the ball is 24.8 Ns — away from the batter (remember the impulse is a vector!) (b) $\mathbf{v}_f = 47.5\,\mathrm{m/s}$ - away from batter.

Answer to Problem: 4.4

$F(t) = 0.090\,\mathrm{N} - (0.12\,\mathrm{N/s})t$

Answer to Problem: 4.5

$p(t) = -125\,\mathrm{Ns}\cos(2\pi\,\mathrm{s}^{-1})t; \quad p(t = 0.5\,\mathrm{s}) = 125\,\mathrm{Ns}; \quad p(t = 1.0\,\mathrm{s}) = -125\,\mathrm{Ns}.$

Answer to Problem: 4.6

$F = -(100\,\mathrm{N} \cdot \mathrm{s}^{1/2})t - 60\,\mathrm{N}$. Note: if there was no friction the driver would need to break in order to keep moving at a constant velocity.

Answer to Problem: 4.7

$v(t) = (0.23\,\mathrm{m/s})\sin(75\,\mathrm{s}^{-1})t; \quad v_{\max} = 0.23\,\mathrm{m/s}; \quad t = 0.042\,\mathrm{s}$

Answer to Problem: 4.8

$F = (15\,\mathrm{N{\cdot}s}^{1/2})t^{-1/2} - (0.015\,\mathrm{N/s}^{1/2})t^{1/2}$

ANSWER TO PROBLEM: 4.9

$$p = (0.74\,\text{N})t + (4.5\,\text{Ns}^{1/2})t^{3/2} + (6.5\,\text{N·s})(e^{(-0.20\text{s}^{-1})t} - 1)$$

The calculus in this problem is more difficult than other problems you have done. Don't be discouraged if the integral was difficult for you.

ANSWER TO PROBLEM: 4.10

$$v_f = 4.5\,\text{m/s}$$

ANSWER TO PROBLEM: 5.1

(a) $a(t) = d^2y(t)/dt^2 = 3a/4\sqrt{t}$.

(b) $F = d(mv)/dt = ma = 3am/4\sqrt{t}$.

ANSWER TO PROBLEM: 5.2

$\mathbf{F}_{total} = \mathbf{N} + \mathbf{F_f} = m\mathbf{a} = md^2\mathbf{r}/dt^2$. Represent the position vector in terms of polar coordinates in the plane. Velocity is tangent to the circle while the acceleration has a tangential and a radial component. Tangential acceleration is a result of the friction force. Radial acceleration is a centripetal acceleration; $v(t) = v_0/(1 + v_0 t/\mu_k)$.

ANSWER TO PROBLEM: 5.3

(a) $c = A\rho = 4 \times 10^{-11}\,\text{g/cm}$.

(b) $M\,dv/dt = -cv^2$ and $M_0 v_0 = Mv$ yield $a = dv/dt = -cv^3/M_0 v_0$.

(c) $t = \frac{M_0 v_0}{2cv_0^2}\left((0.9)^{-2} - 1\right) = 7.73 \times 10^8\,\text{s} \approx 24\,\text{yr}$.

ANSWER TO PROBLEM: 5.4

$F_{net} = 0 = \frac{dp_{rocket}}{dt} + \frac{dp_{fuel}}{dt}$; $\frac{dp_{rocket}}{dt} = \frac{dM}{dt}v + M\frac{dv}{dt} = -\alpha v + Ma$, and $\frac{dp_{fuel}}{dt} = \frac{dm}{dt}v_{fuel} + m\frac{dv_{fuel}}{dt} = -\frac{dM}{dt}v_{fuel} = \alpha(v - V_0)$;
$a = \alpha V_0/M$;
$v(t) = v_0 + V_0 \ln\frac{M_0}{M_0 - \alpha t}$.

ANSWER TO PROBLEM: 5.5

(a) $F_{net} = -Mg + \beta V_0 = M\frac{dv}{dt}$.
When $a = 0$, then, $0 = \beta V_0 - Mg = -\frac{dM}{dt}V_0 - Mg$ so, $M(t) = M_0 e^{-gt/V_0}$ and
$\beta(t) = -\frac{dM}{dt} = \frac{M_0 g}{V_0}e^{-gt/V_0}$.

(b) $M(t) = M_0 - \alpha t$ and $\frac{dv}{dt} = -g + \frac{\alpha V_0}{M_0 - \alpha t}$. With $v_0 = 0$, integration yields $v(t) = -gt + V_0 \ln \frac{M_0}{M_0 - \alpha t}$.

(c) $M(t) = 3M_0/4$ with $\alpha = \frac{2M_0 g}{V_0}$ implies $t = V_0/8g = 167\,\text{s}$. Hence, $v_{3/4} = -g\frac{V_0}{8g} + V_0 \ln \frac{4}{3} = 260\,\text{m/s}$.

ANSWER TO PROBLEM: 5.6

$F = \frac{dm}{dt}v + m\frac{dv}{dt} = \frac{dm}{dt}v - mg$; $\frac{dm}{dt} = -mv/L$; Hence, $F = -mv^2/L - msg/L = -2mgs/L - mgs/L = -3mgs/L$. The force is three times the weight of the portion of the chain already on the platform.

ANSWER TO PROBLEM: 5.7

Use figure EX-5.7a in setting up the problem.

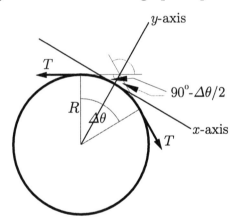

Figure: EX-5.7a

Because of the equilibrium situation, the tangential forces cancel out. The radial force points toward the center and is balanced by the normal force. It equals $F = T\Delta\theta$ for $\Delta\theta << 1$.

Integrating the radial force over the portion of the cylinder in contact with the rope yields: $F_{horizontal} = 0$ and $F_{vertical} = 2T$.

ANSWER TO PROBLEM: 5.8

$v(4\,\mathrm{s}) = 74\,\mathrm{m/s}.$

ANSWER TO PROBLEM: 5.9

$\frac{dv}{dt} = g + cv^2; \; v = -\sqrt{\frac{-g}{c}}\tanh(t\sqrt{-cg});$

$v_{terminal} = -\sqrt{\frac{-g}{c}}$ and is achieved after an infinite amount of time!

ANSWER TO PROBLEM: 5.10

When the length s of the chain is lifted from the table the net force has two contributions – weight $\lambda g s$ of the lifted part and a term that accounts for the change in the momentum of the part that has been elevated, $dp/dt = \lambda v \, ds/dt = \lambda v^2$. Hence, $F_{lift} = \lambda(gs + v^2) = \lambda v(gt + v)$.

Chapter 7   GRAVITY ACCORDING TO NEWTON

ANSWER TO PROBLEM: 7.1

Consider one of the planes (surface mass density σ_1) as the source of the gravitational force that acts on the other plane (surface mass density σ_2). The force on an infinitesimal element of the target plane is: $dF_{dm_2,1} = 2\pi G\sigma_1 dm_2 = 2\pi G\sigma_1\sigma_2 dA_2$. The integral, $\int dF_{dm_2,1} = \infty$, because the plane has infinite area. The force per unit area, $dF_{dm_2,1}/dA_2 = 2\pi G\sigma_1\sigma_2 = dF_{dm_1,2}/dA_1$ is constant.

ANSWER TO PROBLEM: 7.2

Consider one of the rods (linear mass density λ_1) as the source of the gravitational force that acts on the other rod (linear mass density λ_2). The force on an infinitesimal element of the target rod is: $dF_{dm_2,1} = \frac{2G\lambda_1 dm_2}{l} = \frac{2G\lambda_1\lambda_2 dl_2}{l}$. The integral, $\int dF_{dm_2,1} = \infty$, because the rod has infinite length. The force per unit length, $dF_{dm_2,1}/dl_2 = 2G\lambda_1\lambda_2/l = dF_{dm_1,2}/dl_1$ is finite.

ANSWER TO PROBLEM: 7.3

Equation of motion: $m\frac{dv}{dt} = -\frac{2Gm\lambda}{r}$. Multiply the equation of motion by $dr = vdt$ to obtain the form that can be integrated; $\int_{v_0}^0 vdv = -v_0^2/2 = -2G\lambda\int_{r_0}^\infty dr/r = -2G\lambda\ln\frac{\infty}{r_0} = \infty$. Hence, the particle can not escape the gravitational field of the rod.

To get the maximum distance solve the equation: $-\frac{v_0^2}{2} = -2G\lambda\ln(r_{max}/r_0)$ for r_{max}. Hence, $r_{max} = r_0 e^{v_0^2/4G\lambda}$.

ANSWER TO PROBLEM: 7.4

The geometry of the problem is given in Fig. EX-7.4. The force is directed toward the center of the ring and equals to $F = -\frac{GmMl}{(l^2+R^2)^{3/2}}$.

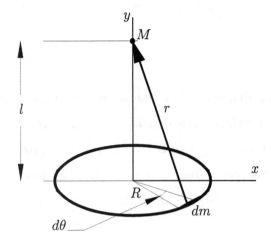

Figure: EX-7.4

ANSWER TO PROBLEM: 7.5

The geometry of the problem is identical to the one depicted in Fig. EX-7.4. The acceleration equation is: $dv/dt = -GMr/(r^2 + R^2)^{3/2}$. The scalar value of the velocity is:

$$v = -\sqrt{\frac{2GM}{R}\left(1 - \frac{R}{\sqrt{r^2+R^2}}\right)}.$$

The maximum distance that the particle can reach on the other side of the ring is $r_{max} = r$.

ANSWER TO PROBLEM: 7.6

The geometry of the problem is depicted in Fig. EX-7.6. Consider one of the spheres as the source of the gravitational force and the other as the target sphere. Imagine the target sphere divided into infinitesimal segments.   The force with which the source sphere acts on the mass element dm of the target sphere is $d\mathbf{F} = -GM_1 dm \frac{x\hat{x}+z\hat{z}+(y+l)\hat{y}}{(x^2+z^2+(y+l)^2)^{3/2}}$. The total force is given by an integral $\int_{target} d\mathbf{F}$. Careful consideration of the integrals yields: $F_x = F_z = 0$ and $F_y = -GM_1 M_2/l^2$.

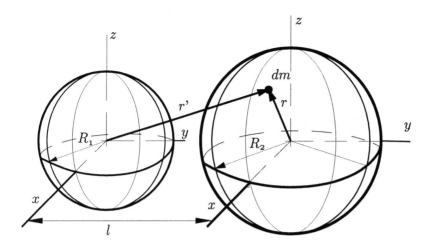

Figure: EX-7.6

ANSWER TO PROBLEM: 7.7

Gravitational force is a centripetal force. For definiteness, assume that the circular orbit of radius R lies, for example, in the x-y plane. Parameterize the orbit by $\mathbf{r} = R(\cos\theta\hat{x} + R\sin\theta\hat{y})$. The acceleration is, $\mathbf{a} = d^2\mathbf{r}/dt^2 = -R\,(d\theta/dt)^2\,(\cos\theta\hat{x} + \sin\theta\hat{y}) = -\,(d\theta/dt)^2\,\mathbf{r}$. Take, $(d\theta/dt)^2 = GM/R^3$. Then, $m\mathbf{a} = -GMm\mathbf{r}/R^3$.

ANSWER TO PROBLEM: 7.8

The equation of motion: $md\mathbf{v}/dt = -GmM_S\mathbf{r}/r^3$. To escape from the Solar system, the satellite must be able to make it to infinity where it will have no velocity. Hence, $v_{escape} = \sqrt{2GM_S/R} = 42 \times 10^4\,\mathrm{m/s}$.

ANSWER TO PROBLEM: 7.9

The gravity of the the Moon and the Earth cancel at the point r_0 obeying the equation (distances are measured form the center of the Earth):
$GM_M/(R_{EM} - r_0)^2 = GM_E/r_0^2$.
Hence, $r_0 = R_{EM}/(1+\sqrt{M_M/M_E}) = 3.47 \times 10^7$ m. Note that r_0 is about $38,000\,\mathrm{km}$ above the surface of the Moon.

The equation of motion is: $m d\mathbf{v}/dt = -GmM_E\mathbf{r}/r^3 - GmM_M\mathbf{r}'/r'^3$. where $\mathbf{r}$ is the position of the space-ship relative to the center of the Earth and $\mathbf{r}'$ is the position of the space-ship relative to the Moon. Since the motion is along the straight line, the following holds: $r + r' = R_{EM}$.

The scalar value of the velocity is:

$$v = \sqrt{2G\left[M_E\left(\frac{1}{R_E} - \frac{1}{r_0}\right) + M_M\left(\frac{1}{R_{EM}-R_E} - \frac{1}{R_{EM}-r_0}\right)\right]}.$$

Numerically, $v = 11,100\,\mathrm{m/s}$.

ANSWER TO PROBLEM: 8.1

$\omega(t) = (3.00\,\text{rad/s}) + (1.95\,\text{rad/s}^3)t^2$

ANSWER TO PROBLEM: 8.2

$\theta(t) = \theta_0 + \omega t = (400\,\text{rad/s})t$

ANSWER TO PROBLEM: 8.3

$\theta(t = 4\,\text{s}) = 2.00$ rotations

ANSWER TO PROBLEM: 8.4

$\omega(t) = 38.7\,\text{rad/s} - 0.667\cos[(12.0\,\text{s}^{-1})t]$

ANSWER TO PROBLEM: 8.5

The child falls down 3.34 seconds after she starts turning. Her angular velocity is 27.6 rad/s.

ANSWER TO PROBLEM: 8.6

The car stops in 7.07 s and has travelled 57.6 m before it stops.

ANSWER TO PROBLEM: 8.7

$I_{\text{total}} = 84.5\,\text{kg·m}^2 + (30\,\text{kg})[1.30\,\text{m} - (0.200\,\text{m/s})t]^2$
The merry-go-round spins faster because the child walked toward the axis of rotation (decreasing I).

ANSWER TO PROBLEM: 8.8

$\alpha = -2.0\,\text{rad/s}^2$

ANSWER TO PROBLEM: 8.9

$\tau = 4.0 \times 10^3\,\text{N·m} + (2.3 \times 10^3\,\text{N·m/s}^2)t^2$

The hamster is ahead of the bottom of the wheel because he is continuously increasing the angular velocity of the wheel.

ANSWER TO PROBLEM: 8.10

$L(t) = 0.022\,\text{kg·m}^2/\text{s} - 0.0017\cos[(3.00\,\text{s}^{-1})t]$

Chapter 9 ENERGY

ANSWER TO PROBLEM: 9.1

T is the tension and F is the pushing force; $F = T\sin\theta = mg\tan\theta$. The work done on the bucket is the work done by the pushing force F,

$W = \int_0^s dx\, F = mg \int_0^s dx \tan\theta = mgl(1 - \cos\theta_f) = 295\,\text{J}$.

The work done by the gravity is $W_G = -(\text{PE}_f - \text{PE}_i) = -mgl(1 - \cos\theta_f) = -295\,\text{J}$.

ANSWER TO PROBLEM: 9.2

It is important that the motion is confined to one-dimension. The force is said to be conservative if the total work performed for a round-trip motion vanishes. $W_{rt} = \int_a^b F(x)dx + \int_b^a F(x)dx = 0$.

ANSWER TO PROBLEM: 9.3

$F(x) = -\frac{\partial\text{PE}(x)}{\partial x} = ax^2 + bx$;

$\text{PE}(x) = -\frac{a}{3}x^3 - \frac{b}{2}x^2 + c = -(1.67\,\text{N/m}^2)x^3 - (1.50\,\text{N/m})x^2$.

ANSWER TO PROBLEM: 9.4

$W = \int_{P_i}^{P_f} \mathbf{F} \cdot d\mathbf{s} = \int_0^{x_0} dx\, ay\big|_{y=0} + \int_0^{y_0} dy\, ax\big|_{x=x_0} = ax_0 y_0$.

$\text{PE} = -axy$.

ANSWER TO PROBLEM: 9.5

$F = -\frac{\partial U(r)}{\partial r} = \frac{2U_0}{b}\left(e^{-2(r-r_0)/b} - e^{-(r-r_0)/b}\right)$. The force is zero at a point $r = r_0 = 0.37 \times 10^{-10}\,\text{m}$.

ANSWER TO PROBLEM: 9.6

(a) $W = \Delta\text{KE} = \frac{mv^2(t)}{2} - \frac{mv_0^2}{2}$; $P = \frac{dW}{dt} = F(t)v(t) = -A + Bt - Ct^2 + Dt^3$,

where $A = mcb = 620\,\text{kW}$, $B = m(2db + c^2) = 252kW/s$, $C = 3mcd = 28\,\text{kW/s}^2$ and $D = 2md^2 = 0.9\,\text{kW/s}^3$.

(b) $KE = \frac{mv^2(t)}{2} = \alpha - \beta t + \gamma t^2 - \delta t^3 + \epsilon t^4$,

where $\alpha = \frac{mb^2}{2} = 1.00\,\text{MJ}$, $\beta = mbc = 0.62MJ/s$, $\gamma = m(2bd+c^2)/2 = 0.126\,\text{MJ/s}^2$, $\delta = mcd = 0.93 \times 10^{-4}\,\text{MJ/s}^3$ and $\epsilon = md^2/2 = 0.225 \times 10^{-3}\,\text{MJ/s}^4$.

$KE(t = 0) = 1.00\,\text{MJ}$; $KE(t = 3\,\text{s}) = 0.27\,\text{MJ}$.

(c) $\Delta KE = KE(t = 3\,\text{s}) - KE(t = 0\,\text{s}) = 0.27\,\text{MJ} - 1.00\,\text{MJ} = -0.73\,\text{MJ}$.

The average power can also be calculated by: $P_{av}(t) = \frac{1}{T} \int_t^{t+T} dt P(t)$.

ANSWER TO PROBLEM: 9.7

The geometry of the problem if given in Fig. EX-9.5.

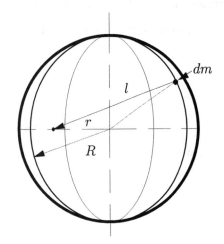

Figure: EX-9.5

$PE = \int_{shell} GM\frac{dm}{l} = \int_0^\pi d\theta \sin\theta \frac{2\pi Gm\frac{M}{4\pi R^2}R^2}{\sqrt{R^2+r^2+2Rr\cos\theta}} = \frac{GMm}{R}$.

ANSWER TO PROBLEM: 9.8

$W = 292\,\text{J}$.

ANSWER TO PROBLEM: 9.9

$x = h(1 - \mu_k \cot\theta)/\mu_k = 1.96\,\text{m}$.

ANSWER TO PROBLEM: 10.1

$m = 0.33\,\text{kg}$

ANSWER TO PROBLEM: 10.2

$\frac{dN}{dt} = 1.50 \times 10^{11}\,\text{molecules/s}$

This is equivalent to $1.50 \times 10^{-8}\,\text{g/hr}$, which seems quite reasonable.

ANSWER TO PROBLEM: 10.3

$\mu_0 = 0.13\,\text{g/cm}^2 = 1.3\,\text{kg/m}^2$

ANSWER TO PROBLEM: 10.4

$t = 100\,\text{s}$. This particular can of whipped cream takes $100\,\text{s}$ to decrease its density to half its original density.

ANSWER TO PROBLEM: 10.5

$\frac{d}{dn}\left(\frac{\Delta V}{V_0}\right) = -6.28 \times 10^{-8}\,\%/\text{m}$

ANSWER TO PROBLEM: 10.6

$\Delta r(h) = -2.48 \times 10^{-9}\,\text{m} - (2.12 \times 10^{-10})h$, where h is in m.

ANSWER TO PROBLEM: 10.7

$\frac{d}{dh}[\rho(h)] = 37.8\,\text{kg/m}^4[1 - (0.014\,\text{m}^{-1})h]^{-2}$

ANSWER TO PROBLEM: 10.8

$W = 0.17\,\text{J}$

ANSWER TO PROBLEM: 10.9

$\sigma = (5.76 \times 10^5\,\text{Pa})(\epsilon)^{1/2}$

ANSWER TO PROBLEM: 10.10

The hanging pizza dough has a length of $1.6\,\text{m} + 2.2 \times 10^{-4}\,\text{m}$.

Chapter 11   FLUIDS

ANSWER TO PROBLEM: 11.1

Use Fig. EG-11.1 as a reference.

$\tau = \int_0^H d\tau = \int_0^H y\rho g(H-y)wdy = \rho gHw\frac{H^2}{2} - \rho gw\frac{H^3}{3} = \frac{1}{6}\rho gwH^3 = 4.4 \times 10^9\,\text{N m}.$

$\tau = \frac{1}{6}\rho gAH^2 = r_{eff}F = r_{eff}\frac{1}{2}\rho gAH.$

$r_{eff} = \frac{H}{3}$; Making the bottom part of the dam heavy (much wider the the top) insures better structural stability!

ANSWER TO PROBLEM: 11.2

Assumptions about the state of the atmosphere: steady and that the air is not at all moving; Bernoulli's equation holds; there is no temperature gradient $-\frac{P}{\rho} =$ constant.

$P_1 - P_2 = -dP = \rho g(y_1 + dy - y_1) = \rho g dy = g\frac{\rho_0}{P_0}Pdy;$

$P = P_0 e^{-\alpha h}.$

ANSWER TO PROBLEM: 11.3

$v = \frac{Q}{A} = \frac{1}{A}\frac{dV}{dt};$

at the point where the pipe diameter is $10\,\text{cm}$, $v = 4.23\,\text{m/s}$;

at the point of the pipe where the diameter is $5\,\text{cm}$. $v = 17.0\,\text{m/s}$; The water flows faster through the narrow portion of the pipe.

ANSWER TO PROBLEM: 11.4

$F = 1.32 \times 10^{11}\,\text{N}.$

ANSWER TO PROBLEM: 11.5

$P_{max} = \frac{dW}{dt} = \frac{dm}{dt}gh = \rho Qgh = 5\,\text{GW};$

Since the turbine is 90% efficient the available power is $P_{eff} = 0.9P_{max} = 4.5\,\text{GW}.$

ANSWER TO PROBLEM: 11.6

$v = \sqrt{2g(H - h)}$; $x = \sqrt{4h(H - h)}$;

From $dx/dh = 0$. the maximum occurs at $h = H/2$;

$x_{max} = x(h = H/2) = H$.

ANSWER TO PROBLEM: 11.7

The geometry of the problem is given in Fig. EX-11.7.

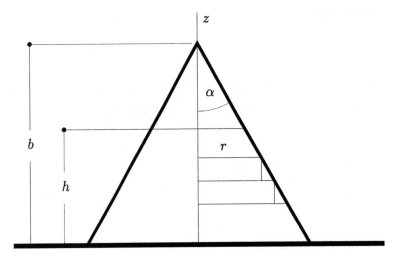

Figure: EX-11.7

The hydrostatic pressure inside the cup is $P(z) = \rho g(h - z)$. because of the conical shape there is unbalanced vertical force $dF = -P(z)dA = -\rho g(h - z)2\pi r dr$.

$F = 2\pi\rho g \tan^2 \alpha \int_0^h dz \, (h - z)(b - z) = \frac{\pi\rho g h^2}{3}(3b - h)$.

ANSWER TO PROBLEM: 11.8

$\frac{dm}{dt} = (2\,\mathrm{m^3/s})\rho_i$; $\rho_i(t) = \rho_0 e^{0.2t}$. Time is measured in seconds!

ANSWER TO PROBLEM: 11.9

$\frac{dm}{dt} = V\frac{d\rho}{dt} = -\rho Av$; $\rho(t) = (16.1\,\mathrm{kg/m^3}) \exp\{-(14.5\,\mathrm{s^{-1}})t\}$.

ANSWER TO PROBLEM: 12.1

$$y(t) = 0.381 \, \text{m} \cos[(39.4 \, \text{rad/s})t + \pi \, \text{rad}] + 0.381 \, \text{m}$$
$$= 0.381 \, \text{m}[1 - \cos(39.4 \, \text{rad/s})t]$$
$$a(t) = -591 \, \text{m/s}^2 \cos[(39.4 \, \text{rad/s})t + \pi \, \text{rad}]$$
$$= 591 \, \text{m/s}^2 \cos[(39.4 \, \text{rad/s})t]$$

ANSWER TO PROBLEM: 12.2

$|A| = 1.59 \, \text{m}$.

ANSWER TO PROBLEM: 12.3

$$v(t) = -(13 \, \text{m/s}) \sin(4.2 \tfrac{\text{rad}}{\text{s}} t)$$

ANSWER TO PROBLEM: 12.4

$$\text{KE}(t) = 1.96 \, \text{J} \sin^2(0.990 \tfrac{\text{rad}}{\text{s}})t$$

ANSWER TO PROBLEM: 12.5

$$v(t) = -\left[(0.45 \, \text{m/s}) \cos\left(4.2 \tfrac{\text{rad}}{\text{s}} t\right) + (6.3 \, \text{m/s}) \sin\left(4.2 \tfrac{\text{rad}}{\text{s}} t\right)\right] e^{-(0.30 \text{s}^{-1})t}$$

KE= 10.6 J. Since the girl's velocity is negative, she jumps off the swing when it is going backwards.

ANSWER TO PROBLEM: 12.6

Initially, as the mass just starts to decrease, the period of the oscillations increases by $2.2 \times 10^{-2} \, \text{sec/kg}$.

ANSWER TO PROBLEM: 12.7

Initially the frequency decreases by $-1.1 \times 10^{-2} \, \text{Hz/m}$ as the length is decreased.

ANSWER TO PROBLEM: 12.8

The total kinetic energy of the wave pulse at $t = 0$ is $0.012\,$J. The kinetic energy of the wave pulse will be 0.012J at any time, assuming the amplitude and frequency are constant.

ANSWER TO PROBLEM: 12.9

PE= $2.0 \times 10^3\,$J and KE= $8.2 \times 10^2\,$J, at $t = 0.70\,$s

ANSWER TO PROBLEM: 12.10

It takes the pulse 0.42s to reach the ceiling.

Chapter 13   SOUND

ANSWER TO PROBLEM: 13.1

Use the analogy with the system described in the Example problem 13.4.

The gas in the cylinder behaves as an elastic spring of spring constant $K_{eff} = -A^2 \left(\frac{\partial P}{\partial V}\right)_0$. The gas density can be written as $\rho = M/V = M/LA$ so, $M/L = A\rho$ plays the role of the linear mass density.

$$v = \sqrt{\frac{K_{eff}L}{\mu}} = \sqrt{\frac{-A^2\left(\frac{\partial P}{\partial V}\right)_0 L}{A\rho}} = \sqrt{-\frac{V\left(\frac{\partial P}{\partial V}\right)_0}{\rho}} = \sqrt{\frac{B}{\rho}}.$$

ANSWER TO PROBLEM: 13.2

$$u_{av} = f \int_0^T dt\, u_0 \cos(kx - \omega t) = 0;$$

$$KE_{av} = f \int_0^T dt\, KE = \frac{\Delta V f \rho \omega^2 u_0^2}{2} \int_0^T dt\, \sin^2(kx - \omega t) = \Delta V \frac{\omega^2 \rho u_0^2}{4}.$$

$$PE_{av} = f \int_0^T dt\, PE = \frac{\Delta V f \rho \omega^2 k^2 u_0^2}{2} \int_0^T dt\, \sin^2(kx - \omega t) = \Delta V \frac{\rho \omega^2 u_0^2}{4}.$$

Note that the average kinetic and potential energies are equal.

ANSWER TO PROBLEM: 13.3

Doppler effect; $f(t) = \frac{f_0 v_s}{at + v_s}$;

$$\Delta f = \frac{df}{dt} \Delta t = f(t)\left(1 - \frac{v_s}{at + v_s}\right).$$

ANSWER TO PROBLEM: 13.4

(a) $PV = nRT$; $PV^\gamma = $ constant; $B = -V\frac{\partial P}{\partial V} = \frac{\gamma P}{V}$. $v = \sqrt{\frac{\gamma P}{\rho}}$; $\rho = mP/RT$; $f = v/\lambda = \sqrt{m\gamma RT}$; $f/f_0 = \sqrt{T/T_0}$.

$$\frac{\Delta f}{f_0} = \frac{\Delta T}{2\sqrt{TT_0}} = \frac{at}{2\sqrt{T_0(T_0 + at)}}.$$

(b) $\frac{\Delta f}{f_0} = 0.044$ and $f = f_0 + \Delta f = 530\,\text{Hz}$.

ANSWER TO PROBLEM: 13.5

$$\Delta P = -B\frac{\partial s(x,t)}{\partial x} = -v^2 \rho \frac{\partial s(x,t)}{\partial x} = v^2 \rho s_0 k \sin(kx - \omega t).$$

$$\Delta P_{av} = 0.$$

ANSWER TO PROBLEM: 13.6

$E_{av} = KE_{av} + PE_{av} = \Delta V \frac{\rho \omega^2 s_0^2}{2}$.

The average power is the amount of energy delivered per unit time. $P_{av} = \frac{E_{av}}{\Delta t} = \frac{A\rho\omega^2 s_0^2}{2} \frac{\Delta x}{\Delta t} = \frac{1}{2} A\rho v \omega^2 s_0^2$.

The intensity is the amount of power delivered per unit area. $I = P_{av}/A = \rho\omega^2 s_0^2/2$.

Chapter 14   THERMAL PROPERTIES OF MATTER

ANSWER TO PROBLEM: 14.1

$L_f = 0.3504 \, \mathrm{m}$. The material was longer by $0.4 \, \mathrm{mm}$ because of being heated.

ANSWER TO PROBLEM: 14.2

$\beta(t) = 25 \times 10^{-6} \, \mathrm{K}^{-1} + (3.3 \times 10^{-8} \, \mathrm{K}^{-3}) T^2$

ANSWER TO PROBLEM: 14.3

$\alpha(T) = (8.6 \times 10^{-6} \, \mathrm{K}^{-1}) + (7.3 \times 10^{-8} \, \mathrm{K}^{-1.7}) T^{0.7}$

ANSWER TO PROBLEM: 14.4

$V(t) = (5.15 \, \mathrm{m}^3) \mathrm{e}^{-(0.100 \mathrm{s}^{-1}) t}$

ANSWER TO PROBLEM: 14.5

$\frac{\partial V}{\partial t} = 2.5 \times 10^{-4} \, \mathrm{m}^3/\mathrm{s}$

ANSWER TO PROBLEM: 14.6

$\frac{\partial T}{\partial t} = -983 \frac{\mathrm{K}}{\mathrm{s}} - 43.3 \frac{\mathrm{K}}{\mathrm{s}^{3/2}} t^{1/2} + 192 \frac{\mathrm{K}}{\mathrm{s}^{1/2}} t^{-1/2}$

ANSWER TO PROBLEM: 14.7

$\left. \frac{\partial V}{\partial t} \right|_{t=0} = 6.3 \times 10^{-2} \frac{\mathrm{m}^3}{\mathrm{s}}$

ANSWER TO PROBLEM: 14.8

$\frac{\partial T}{\partial t} = -0.4 \frac{\mathrm{P}}{\mathrm{nR}} \frac{\partial V}{\partial t}$. Notice that the temperature decreases as the volume increases for an adiabatic process.

ANSWER TO PROBLEM: 14.9

$\beta(T) = \frac{-(A-2BT+3CT^2)}{(1+AT-BT^2+CT^3)}$ where $A = 5.3 \times 10^{-5}\,^{\circ}C^{-1}$, $B = 6.5 \times 10^{-6}\,^{\circ}C^{-2}$, and
$C = 1.4 \times 10^{-8}\,^{\circ}C^{-3}$.

ANSWER TO PROBLEM: 14.10

$F_a = +0.18\,\text{N}$ at $t = 3.0\,\text{hrs}$

Chapter 15   HEAT AND THERMAL ENERGY

ANSWER TO PROBLEM: 15.1

$\frac{dQ}{dt} = 2\pi k L \frac{T-T_o}{\ln\frac{R_o}{R_i}} = (0.032\,\text{cal/s})(T-20)$ where the temperature is measured in degrees Celsius.

$\frac{dQ}{dt} = -mc\frac{dT}{dt} = (0.032\,\text{cal/s})(T-20);\ \frac{dT}{dt} = -\frac{1}{3.12\times10^4\,\text{s}^{-1}}(T-20);$

$\Delta t = -(3.12\times10^4\,\text{s}^{-1})\int_{90}^{50}\frac{dT}{T-20} = 2.65\times10^4\,\text{s} = 441.$

ANSWER TO PROBLEM: 15.2

$\frac{dQ}{dt} = -4\pi k r^2 \frac{dT}{dr};$

$\frac{dQ}{dt} = \frac{4\pi(T_2-T_1)}{\frac{1}{R_2}-\frac{1}{R_1}}.$

ANSWER TO PROBLEM: 15.3

$Q_A + Q_B = \int_{T_A}^{T_f} m_A c_A(T)dT + \int_{T_B}^{T_f} m_B c_B(T)dT = 0.$

$T_f = \frac{m_A c_A T_A + m_B c_B T_B}{m_A c_A + m_B c_B}.$

ANSWER TO PROBLEM: 15.4

$dQ = mcdT;\ Q = 1.18\times10^5\,\text{cal}.$

ANSWER TO PROBLEM: 15.5

$dU = dQ = \frac{3}{2}nRdT;\ n = m/m_{mol};\ c = \frac{1}{m}\frac{dQ}{dT} = \frac{m_{mol}}{n}\frac{3}{2}nR = 3m_{mol}R/2;$

Specific heat per mole at constant volume is $3R/2$.

$dU = dQ - PdV;\ PdV = nRdT;\ dQ = \frac{3}{2}nRdT + nRdT = \frac{5}{2}nRdT;$

The specific heat per mole at constant pressure is $5R/2$.

$c_P - c_V = R.$

ANSWER TO PROBLEM: 15.6

$Q_l + Q_h = \frac{mk}{4}(T_f^4 - T_l^4) + \frac{mk}{4}(T_f^4 - T_h^4) = 0.\ T_f = \left(\frac{T_l^4 + T_h^4}{2}\right)^{1/4}.$

$dQ/dt = -kA dT/dy$; $Q = mL$ and $m = \rho A y$.

$dQ/dt = L dm/dt = -kA dT/dy = -kA(-T - 0)/y = kAT/y$; $dm/dt = \rho A dy/dt$.

$dy/ty = kT/\rho L y$.

ANSWER TO PROBLEM: 16.1

$T_2 = 220\,\text{K}$; $W_{\text{needed}} = 52.5\,\text{J}$

ANSWER TO PROBLEM: 16.2

$W = 14\,\text{kJ}$

ANSWER TO PROBLEM: 16.3

$W = 1.5 \times 10^2\,\text{J}$

ANSWER TO PROBLEM: 16.4

$W = 17.9\,\text{kJ}$

ANSWER TO PROBLEM: 16.5

$T = (1.09\,\text{K/m}^3)V + (0.437\,\text{K/m}^6)V^2$ where V is in units of m^3.

ANSWER TO PROBLEM: 16.6

$W = 43.2\,\text{kJ}$

ANSWER TO PROBLEM: 16.7

$\Delta U = 1.73 \times 10^3\,\text{kJ}$

ANSWER TO PROBLEM: 16/8

$\left|\frac{de_c}{dT_L}\right| > \left|\frac{de_c}{dT_H}\right|$, so it is better to lower T_L than to raise T_H if you want the largest increase in efficiency.

ANSWER TO PROBLEM: 16.9

$\Delta S = -5.72\,\text{kJ/K}$

ANSWER TO PROBLEM: 16.10

$\Delta S = 28\,\text{J/K}$

ANSWER TO PROBLEM: 17.1

The electric field of the semi-ring at the center of the circle has the vertical component only: $E = \frac{2\lambda}{4\pi\epsilon_0} r$.
$F = qE = \frac{2q\lambda}{4\pi\epsilon_0} r = 18 \times 10^{-3}\,\mathrm{N}$.
The direction of the force is the same as the direction of the electric field.

ANSWER TO PROBLEM: 17.2

Cube has six sides but non-zero electric flux is only through the faces perpendicular
to x-direction, the direction of the electric field.
$\Phi = \int_{face} \mathbf{E} \cdot \hat{x}\,dxdy + \int_{face} \mathbf{E} \cdot (-\hat{x})dxdy = 0$, because the electric field is uniform.

ANSWER TO PROBLEM: 17.3

See Fig. EX-17.3. Select the coordinate system such that the origin is at the center
of the flat face of the hemisphere and the z-axis points away from the hemisphere,
perpendicular to the flat face. The x and y axis are in the plane of the flat face.
The electric field of the charge Q is $\mathbf{E} = \frac{Q}{4\pi\epsilon_0} \frac{\mathbf{r}-\delta\hat{z}}{|\mathbf{r}-\delta\hat{z}|^3}$.
(a) $\Phi_{ff} = -\frac{Q}{2\epsilon_0}\left[\frac{1}{\sqrt{1+(R/\delta)^2}} - 1\right]$.
(b) By Gauss's theorem, net flux vanishes, hence, $\Phi_{cs} = -\Phi_{ff}$.

ANSWER TO PROBLEM: 17.4

To visualize the geometry draw the picture. $\Phi = \int_{triangle} \mathbf{E} \cdot d\mathbf{A} = bwA$, where
$A = hW/2$ is the area of the triangle.

ANSWER TO PROBLEM: 17.5

Only the surfaces perpendicular to the direction of the electric field contribute to
the flux.
(a) $\Phi = \int_a^{a+c} dx \int_0^a dy(3 + 2x^2) = 0.97\,\mathrm{Nm^2/C}$.

(b) By Gauss's law, $Q = \epsilon_0 \Phi = 8.6 \times 10^{-12}\,\text{C}$.

Answer to Problem: 17.6

The electric field inside the charge distribution: $\mathbf{E} = \frac{1}{4\pi\epsilon_0}\frac{q\mathbf{r}}{R^3}$. The force on the electron at a location $\mathbf{r}$ relative to the center of the distribution is $\mathbf{F} = -\frac{1}{4\pi\epsilon_0}\frac{q^2\mathbf{r}}{R^3}$. Applying Newton's second law yields an equation of motion for the electron, $m\mathbf{a} = \mathbf{F} = -\frac{1}{4\pi\epsilon_0}\frac{q^2\mathbf{r}}{R^3}$. Therefore, $\frac{d^2r}{dt^2} + \frac{1}{4\pi\epsilon_0}\frac{q^2}{R^3}r = 0$, which is the equation of the harmonic oscillator with frequency $f = \frac{1}{2\pi}\sqrt{\frac{1}{4\pi\epsilon_0}\frac{q^2}{mR^3}} = 6.6 \times 10^{15}\,\text{Hz}$.

Answer to Problem: 17.7

$E(t) = at$; $m\frac{d^2x}{dt^2} = -qE = eat$; $v = \frac{ea}{2m}t^2$; $x = \frac{ea}{6m}t^3$.

(a) $x(0.1\,\text{ns}) = 0.177\,\text{mm}$.

(b) The total time of flight between the cathode and the anode divides into two parts - time interval during which the electron experienced the acceleration and the time interval during which it moved at constant speed.

acceleration: $\Delta t = 0.1\,\text{ns}$. At the moment the acceleration stopped the speed was $v = 3.52 \times 10^4\,\text{m/s}$; at that speed the electron covered the distance, $d - x$; the total time of flight is $t = 1.427\,\mu\text{s}$.

Answer to Problem: 17.8

Electric field is radial. (a) In the interior of the sphere: $E = \frac{Q}{4\pi\epsilon_0}\frac{r^3}{R^5}$. (b) Outside the sphere: $E = \frac{1}{4\pi\epsilon_0}\frac{Q}{r^2}$.

Answer to Problem: 17.9

For practical purposes the tube can be considered as infinite in length. Use Gauss's law.

For $r < a$, $E(r < a) = 0$.

For $a < r < b$ $E(a < r < b) = \frac{Q/L}{2\pi\epsilon_0}\frac{r^2-a^2}{b^2-a^2}$.

For $r > b$, $E(r > b) = \frac{Q/L}{2\pi\epsilon_0 r}$.

Note that the electric filed is a continuous function of the position coordinate r.

From: $-e = \int_0^\infty \rho(r)dV$, follows, $C = e/\pi a_0^3$.

$\rho = -\frac{e}{\pi a_0^3} e^{-2r/a_0}$; $Q_{a_0} = \int_0^{a_0} \left(-\frac{e}{\pi a_0^3}\right) e^{-2r/a_0} 4\pi r^2 dr = -0.32e$.

$E_{a_0} = \frac{0.32e}{4\pi\epsilon_0 a_0^2} = 1.6 \times 10^{11}$ N/C.

Chapter 18   Electrostatics: Energy

ANSWER TO PROBLEM: 18.1

$E_x = -17.3 \, \text{V/m}^3$; $E_y = -5.40 \, \text{V/m}^2 y$; $E_z = -7.37 \, \text{V/m}$.
$|E| = 124 \, \text{V/m}$ at $(7.00 \, \text{m}, -5.00 \, \text{m}, \ 3.00 \, \text{m})$

ANSWER TO PROBLEM: 18.2

$E_x = -7.15 \, \text{V/m}^3 yz$; $E_y = -(7.15 \, \text{V/m}^3 xz + 2.45 \, \text{V/m}^2 z)$; $E_z = -(7.15 \, \text{V/m}^3 xy +$
$2.45 \, \text{V/m}^2 y)$
$|E| = 25.8 \, \text{V/m}$ at $(1.00 \, \text{m}, -1.00 \, \text{m}, \ 2.00 \, \text{m})$.

ANSWER TO PROBLEM: 18.3

$x_0 = 10.5 \, \text{m}$; $v = 6.67 \times 10^6 \, \text{m/s}$. If the electron starts at $x = 10.5 \, \text{m}$ and travels
at the same speed as the potential, it will never feel a force due to the potential.

ANSWER TO PROBLEM: 18.4

$\Delta y = +95.8 \, \text{m}$. After the proton has moved $2.0 \, \text{m}$ in the $+x$-direction, it has moved
$95.8 \, \text{m}$ in the positive y-direction.

ANSWER TO PROBLEM: 18.5

$\Delta V = +679 \, \text{V}$. The potential is higher at $(7.00 \, \text{m}, \ 3.00 \, \text{m})$ than at $(5.00 \, \text{m}, \ 5.00 \, \text{m})$.

ANSWER TO PROBLEM: 18.6

$V(r = 3.00 \, \text{m}) = 6.50 \times 10^8 \, \text{J/C}$. Notice that this is the exact same potential you
would get if all the charge was concentrated at the center of the sphere.

ANSWER TO PROBLEM: 18.7

$V(z) = (1.13 \times 10^5 \, \text{N/C})(\sqrt{z^2 + 100 \, \text{m}^2} - z)$

ANSWER TO PROBLEM: 18.8

$W = \frac{1}{2}\frac{kQ^2}{R}$

ANSWER TO PROBLEM: 18.9

$E_x = -(20.0\,\text{V/mm})$. The plate ar $x = 25.0\,\text{mm}$ is positively charged and the plate at $x = 0.0\,\text{mm}$ is negatively charged.

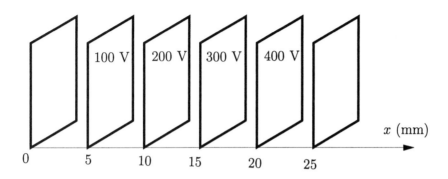

Figure: EX-18.9a

ANSWER TO PROBLEM: 18.10

$37.5\,\mu\text{F}$. Notice, if the dielectric completely filled the space then $C_{new} = K_e C$.

ANSWER TO PROBLEM: 19.1

$I = dq/dt = (6.00\,\text{C/s}^2)t^2 + (12.00\,\text{C/s})t.$

ANSWER TO PROBLEM: 19.2

$I = \int J dA = \int_0^R J(r)2\pi r dr = 2J_0\pi R^2/3.$

ANSWER TO PROBLEM: 19.3

The current of the positive charges is a result of the motion of the nuclei of oxygen and hydrogen molecules. $dN/dt = Q\rho N_A/M$ is the rate of molecule flow through the hose. There are $8 + 2 = 10$ positive charges per water molecule.
$I_+ = dQ/dt = 10edN/dt = 10eR\rho N_A/M = 24 \times 10^6$ A.
The current of the positive charges is compensated by the current of the electrons which flow in the same direction (along with molecules) and yield the current of the same magnitude but of the opposite signature.

ANSWER TO PROBLEM: 19.4

$dR = \rho dx/\pi r^2.$ $R = \int_0^l dR = \frac{\rho l}{a(b-a)}\int_a^b \frac{dr}{r^2} = \rho a l/\pi a^2 b.$

EXERCISE   19.5

$I = dq/dt;$ $J = I/A = \sigma E;$ $\sigma = 1/\rho$ and $E = -dV/dx.$
$I = -(A/\rho)dV/dx = -(A/\rho)kV_0 \cos kx = (5.0\,\text{A}) \cos kx.$

ANSWER TO PROBLEM: 19.6

The electron moves under the influence of the electric field produced by two walls of charge, $E = -Q/A\epsilon_0$ where A is the area of the charge wall. The charges on the walls are due to the electrons that are displaced from the equilibrium positions. Hence, if there are n electrons per unit volume and each is displaced by distance x. then the total charge on the wall is $Q = nqAx$, where q is the electron charge.

The force on the electron is given by Newton's second law, $md^2x/dt^2 = qE$, where m is the electron mass. Combining the expression for the displaced charge and the expression for the induced electric field with the acceleration equation, yields an equation for the electric charge, $d^2Q/dt^2 = -nq^2Q/m\epsilon_0$. Using, $I = dQ/dt$ yields a current equation: $d^2I/dt^2 = -nq^2I/m\epsilon_0$. Since this is a harmonic oscillator equation it means that the electron current in the ionosphere oscillates with frequency $f = \frac{1}{2\pi}\sqrt{nq^2/\epsilon_0 m}$. Using the numerical value for the observed daytime frequency yields that there are 10^6 to 10^7 electrons per cubic centimeter in the ionosphere!

ANSWER TO PROBLEM: 19.7

$I = dQ/dt = \frac{Q_0}{\tau}e^{-t/\tau}$. During the time interval of length τ the current drops by about 67%, that is to 37% of the initial value. τ is known as the time characteristic of the system

ANSWER TO PROBLEM: 19.8

$I = \int_{area} J(r)dA = 2\pi J_0 \int_0^R (1 - e^{-r/\delta})rdr = 2\pi J_0 \left[\frac{R^2}{2} + \delta(e^{-R/\delta} - 1)\right]$.

Chapter 20   CIRCUITS

ANSWER TO PROBLEM: 20.1

$Q(t = 2.00\,\text{s}) = 1.77\,\mu\text{C}$

ANSWER TO PROBLEM: 20.2

$I(t = 215\,\text{ms}) = 0.109\,\text{A}$

ANSWER TO PROBLEM: 20.3

The total heat generated at the resistor is 1.84 J. Notice that this is equal to the total energy stored in the capacitor.

ANSWER TO PROBLEM: 20.4

$I(t = 0.425\,\text{s}) = 15.9\,\text{mA}$

ANSWER TO PROBLEM: 20.5

$W = 1.68\,\text{J}$ of heat dissipated. When trying to charge a capacitor, as much energy is lost to heating as is stored in the capacitor.

ANSWER TO PROBLEM: 20.6

$t = 1.84\,\text{min}$. Extra Practice: Try this problem using Kirchoff's rules to get the same answer.

ANSWER TO PROBLEM: 20.7

$t = 4.72\,\text{min}$

ANSWER TO PROBLEM: 20.8

$t = 1.18\,\text{min}$

ANSWER TO PROBLEM: 20.9

$t = 12.4$ minutes. Notice, however, that a capacitor is not the best element to store the energy of the solar cell. Extra thought: Can you think of a way to build a "battery" of capacitors that would lengthen the time the lightbulb is lit?

ANSWER TO PROBLEM: 20.10

$$I_1(t) = 20.7 \, \text{mA}(1 - e^{-t/(37.8\text{s})})$$

Chapter 21   MAGNETISM

ANSWER TO PROBLEM: 21.1

$B = \frac{\mu_0 I r}{2R^2 \pi}$;

$E_B = \frac{1}{2\mu_0} \int \mathbf{B}^2 dV = \mu_0 I^2 L / 16\pi = 2.5 \times 10^{-6}\,\text{J}.$

ANSWER TO PROBLEM: 21.2

I is the current in the inner conductor; $-I$ is the current in the outer conductor. The direction is determined using the right hand rule.

For $r < R_1$, $B = \frac{\mu_0 I}{2\pi R_1^2} r$;

For $R_1 < r < R_2$, $B = \frac{\mu_0 I}{2\pi r}$;

For $R_2 < r < R_3$, $B = \frac{\mu_0 I}{2\pi} \frac{R_3^2 - r^2}{r(R_3^2 - R_2^2)}$;

For $r > R_2$, $B = 0$.

The magnetic field is a continuous function of distance.

ANSWER TO PROBLEM: 21.3

b is the vertical distance from the observation point to the straight line. θ_1 and θ_2 are the angles at which the ends of the cable are seen from the observation point.

$B = \frac{\mu_0 I}{4\pi b}(\sin\theta_2 - \sin\theta_1)$.

ANSWER TO PROBLEM: 21.4

The magnetic field points perpendicular to the plane of the circle; $B = \frac{\mu_0 I L}{4\pi R^2}$.

ANSWER TO PROBLEM: 21.5

The magnetic field is tangential to circles concentric with the axis of the conducting wire.

For $0 \leq r \leq R$, $B = \mu_0 J \delta \left(\frac{\delta}{r}(1 - e^{-r/\delta}) - e^{-r/\delta} \right)$;

For $r > R$, $B = \frac{\mu_0 J_0 \delta}{2\pi r} \left(\delta(1 - e^{-R/\delta}) - R e^{-R/\delta} \right)$.

(Useful formula from the integral calculus: $\int u \; dv = uv - \int v \; du$.)

ANSWER TO PROBLEM: 21.6

The equation of motion: $\mathbf{F} = q\mathbf{v} \times \mathbf{B} = md\mathbf{v}/dt$. $\mathbf{B} = B\hat{z}$.

$mdv_x/dt = qv_yB$, $mdv_y/dt = -qv_xB$, and $mdv_z/dt = 0$.

$v_z = v_{z0} = 0$; $v_x = v_0 \cos \omega t$; $v_y = -v_0 \sin \omega t$.

$\mathbf{v}^2 = \mathbf{v}_0^2$; $\omega = qB/m$; $R = mv_0/qB$.

ANSWER TO PROBLEM: 21.7

$W = \int_{path} q(\mathbf{v} \times \mathbf{B}) \cdot d\mathbf{l} = q \int_{path} (\mathbf{v} \times \mathbf{B}) \cdot \mathbf{v}dt = 0$.

ANSWER TO PROBLEM: 21.8

$d\mathbf{F} = Id\mathbf{l} \times \mathbf{B}$. Vector $d\mathbf{l}$ is tangent to the current loop. Vector $d\mathbf{l} \times \mathbf{B}$ is in the radial direction. Hence, $\mathbf{F} = \int_{loop} d\mathbf{F} = 0$.

EXERCISE 21.9

Semicircular loop of radius R: $F = 2IBR$.

Straight piece of wire of length $2R$: $F = 2RIB$.

The direction of the two forces are the same which follows by applying the right hand rule.

Chapter 22   ELECTROMAGNETIC INDUCTION

ANSWER TO PROBLEM: 22.1

$$\Phi_B = 8.0 \times 10^{-9} Tm^2 \ln\left[\frac{1.2\text{m} + 0.75\text{m}\cos(60\,\text{rad/s}\,t)}{1.2\text{m} - 0.75\text{m}\cos(60\,\text{rad/s}\,t)}\right]$$

ANSWER TO PROBLEM: 22.2

$\varepsilon = 8.64\,\text{V}\sin(6\pi\frac{\text{rad}}{\text{s}})t$

ANSWER TO PROBLEM: 22.3

$$\varepsilon(t) = 2.98\,\text{V}\left[\sin\left(\frac{2\pi\text{rad}}{\text{s}}t\right) + \sin\left(\frac{2\pi\text{rad}}{\text{s}}t + \frac{2\pi}{3}\,\text{rad}\right)\right]$$

ANSWER TO PROBLEM: 22.4

$\varepsilon = 2.80\,\text{mV}$. Even though B is changing and the wire is moving, the emf in the loop is constant.

ANSWER TO PROBLEM: 22.5

$P_2(t) = (9.5 \times 10^{-11}\,\Omega\text{s}^2)[(0.30\,\text{A/s}^2)t - 2.1\,\text{A/s}]^2$

ANSWER TO PROBLEM: 22.6

$\varepsilon(t) = -8.2 \times 10^{-2}\,\text{V} - (1.4 \times 10^{-2}\,\text{V/s})t$

ANSWER TO PROBLEM: 22.7

$\varepsilon(t) = 0.90\,\text{V} - \left(0.090\frac{\text{V}}{\text{s}}\right)t$ for $0 < t < 0.1\,\text{s}$.

ANSWER TO PROBLEM: 22.8

$I = 12.0\,\text{mA}$. The current generated is proportional to the acceleration of the plane.

ANSWER TO PROBLEM: 22.9

$I = 3.79\,\text{A}$. Note: at $t = \infty$, $I = 8.0\,\text{A}$.

ANSWER TO PROBLEM: 22.10

$P = 1.35W\text{e}^{-(150\text{s}^{-1})t}$

EXERCISE 23.1

The circuit equation is that of an LRC circuit with all elements connected in series,
$\frac{d^2V}{dt^2} + \frac{R}{L}\frac{dV}{dt} + \frac{1}{LC}V = 0$. The assumed solution is of the form $V = V_0 e^{-\beta t}$;
$\beta^2 - \beta R/L + 1/LC = 0$; $\beta_{1,2} = \frac{R}{2L}\left(1 \pm \sqrt{1 - \frac{4L}{R^2 C}}\right)$; $\beta_1 = 5.84 \times 10^6\,\text{s}^{-1}$ and
$\beta_2 = 0.171 \times 10^6\,\text{s}^{-1}$.
$V = Ae^{-\beta_1 t} + Be^{-\beta_2 t}$; At $t = 0$, $I = 0$ or equivalently, $dV/dt = 0$; $A/B = -\beta_2/\beta_1 = -1/34$.
The term $Be^{-\beta_1 t}$ decays fast so the terms $Ae^{-\beta_2 t}$ is dominant after about a microsecond.

ANSWER TO PROBLEM: 23.2

$V = Ae^{-\alpha t}\cos\omega t$, $I = ACe^{-\alpha t}(\omega\sin\omega t + \alpha\cos\omega t)$, $\alpha = R/2L$, $\omega^2 = -\alpha^2 + 1/LC$.
$E = \frac{CV^2}{2} + \frac{LI^2}{2} = \frac{CA^2}{2}e^{-2\alpha t}\left[1 + LC(\alpha^2\cos 2\omega t + \alpha\omega\sin 2\omega t)\right]$.
The energy is dissipated because of the exponential term. To dissipate the energy as quickly as possible, α has to be as large as possible.
At $R = 2\sqrt{L/C}$, $\alpha = R/2L = 1/\sqrt{LC}$ and $\omega = 0$; $E = CA^2 e^{-2\alpha t}$ and it drops faster than in the under-damped case. In the overdamped case the solution is of the form $Ae^{-\beta_1 t} + Be^{-\beta_2 t}$ with $\beta_1 = (R/2L)(1 + \sqrt{1 - (4L/R^2 C)})$ and $\beta_2 = (R/2L)(1 - \sqrt{1 - (4L/R^2 C)})$. Since $\beta_2 < \beta_1$ the dominant term is $Be^{-\beta_2 t}$. But since $\beta_2 < R/2L$ the decay is slower than in the critical case.

EXERCISE 23.3

$LdI/dt + RI = V_0\cos\omega t$. In a steady state (after the transient solution has died off) the current oscillates at the frequency of the driving electromotive force: $I = I_0\cos(\omega t + \phi)$; $\tan\phi = -\omega L/R$ and $I_0 = V_0/\sqrt{R^2 + \omega^2 L^2}$.

EXERCISE 23.4

$-Q/C + RI = V_0 \cos \omega t$. In the steady state after the transient solution dies off, the current oscillates at the frequency of the driving electromotive force: $I = I_0 \cos(\omega t + \phi)$; $Q = -\int I dt = -(I_0/\omega) \sin(\omega t + \phi)$; $\tan \phi = 1/R\omega C$; $I_0 = V_0 \sqrt{R^2 + (1/\omega C)^2}$. The current in the circuit leads the voltage of the driving electromotive force.

EXERCISE 23.5

The instantaneous power: $P = VI = V_0 I_0 \cos \omega t \sin(\omega t + \phi)$. The average power given by integral of the instantaneous power over a cycle, $T = 2\pi/\omega$.

$P_{av} = (1/T) \int_0^T P(t)dt = \frac{1}{2} V_0 I_0 \cos \phi$. Note that the average power may be positive or negative, depending on value taken of $\cos \phi$.

(The following integral formulas are useful:

$(1/T) \int_0^T \cos^2 \omega t dt = (1/T) \int_0^T \sin^2 \omega t dt = 1/2$; $(1/T) \int_0^T \cos \omega t \sin \omega t dt = 0$,)

ANSWER TO PROBLEM: 23.6

$\mathcal{E} = Q_2/C = RI_1 + LdI_2/dt$, $I = I_1 + I_2$, $I_2 = dQ_2/dt$;

$RI + L\frac{dI}{dt} = \mathcal{E} - RC\frac{d\mathcal{E}}{dt} - LC\frac{d^2\mathcal{E}}{dt^2}$.

$I = I_0 \cos(\omega t + \phi)$; $V_0(1 - LC\omega^2) = I_0(R\cos \phi - \omega L \sin \phi)$ and $-V_0 RC\omega = -I_0(R\sin \phi + \omega L \cos \phi)$.

The phase $\phi = 0$ if $\omega^2 = 1/LC - R^2/L^2$.

EXERCISE 23.7

$V_{rms} = \sqrt{V_{av}^2}$; $V_{av}^2 = \frac{1}{T} \int_0^T V^2(t)dt$, where $T = 1/f$ and $V(t) = V_0 \cos \omega t$, $\omega = 2\pi f = 2\pi/T$.

$V_0 = \sqrt{2}V_0 = 170\,\text{V}$.

EXERCISE 23.8

$V_0 = \sqrt{2}V_{rms} = 311\,\text{V}$.

283

ANSWER TO PROBLEM: 24.1

The displacement current is equal to the current in the loop.

ANSWER TO PROBLEM: 24.2

$I_D = -\frac{Q_0}{RC}e^{-t/RC}$. Note: the displacement current is flowing counterclockwise.

ANSWER TO PROBLEM: 24.3

$I_D = \varepsilon\frac{d\Phi_E}{dt}$ in any medium

ANSWER TO PROBLEM: 24.4

$B = \frac{\mu_0 I}{2\pi r}$

This may seem surprising, but since there is a displacement current between the plates of the capacitor that equals the current flowing to the capacitor, B is unchanged across the capacitor.

ANSWER TO PROBLEM: 24.5

$B = \frac{\mu_0\varepsilon}{2\pi rR}e^{-t/RC}$

ANSWER TO PROBLEM: 24.6

$B_{ind} = 8.34 \times 10^{-8}\,\text{T}$ (counterclockwise)

ANSWER TO PROBLEM: 24.7

irradiance$= 1.30 \times 10^3\,\frac{N}{\text{m·s}}$

ANSWER TO PROBLEM: 24.8

The total irradiance is $1.26 \times 10^{-10}\,\frac{W}{\text{m}^2}$.

ANSWER TO PROBLEM: 24.9

$P_{AV} = 4.69 \times 10^{-3} W$

ANSWER TO PROBLEM: 24.10

$v(1.2 \times 10^5 \, \text{km}) = 2.42 \, \text{km/s}$

ANSWER TO PROBLEM: 28.1

$KE_{rel} = \frac{1}{2}mv^2$ (if $\frac{v^2}{c^2}$ is small).

ANSWER TO PROBLEM: 28.2

$\Delta t_m = \Delta t_s(3.12 \times 10^{-3})$ at 0.9500c. Compare that to a dilation of 2.75×10^{-5} for 0.2500c → 0.2501c. So, the time dilates more as the object moves faster.

ANSWER TO PROBLEM: 28.3

$\Delta L_m = L_s(-3.04 \times 10^{-4})$ at 0.9500c. Compare this to a contraction of -2.85×10^{-5} for 0.2500c → 0.2501c. So, the length contracts more as the object moves faster.

ANSWER TO PROBLEM: 28.4

$F = m\gamma^3 a.$

ANSWER TO PROBLEM: 30.1

$E = 6.54 \times 10^{-5}$ J.

ANSWER TO PROBLEM: 302.

$\theta = \pi$ gives maximum $\Delta \lambda$.

ANSWER TO PROBLEM: 32.1

$R_f = 1.2 \times 10^{12}$ decay/s.

ANSWER TO PROBLEM: 32.2

$R = 2.9 \times 10^{10}$ decays/day

ANSWER TO PROBLEM: 32.3

$N_0 = 6.93 \times 10^{10}$ atoms.

ANSWER TO PROBLEM: 32.4

$t_{1/2} = 1.71$ hr.

ANSWER TO PROBLEM: 32.5

$t_{1/2} = 18.0$ min

ANSWER TO PROBLEM: 32.6

$R_0 = 1.2 \times 10^{16}$ decays/hr.

ANSWER TO PROBLEM: 32.7

$t_f = 5.21 \times 10^8$ y.

ANSWER TO PROBLEM: 32.8

$t = 2.17 \times 10^3$ y

$$t_{1/2} = \sqrt{\frac{1.386}{B}}.$$

In this case, $t_{1/2}$ is a constant but the dependence of N on time is different than the experimentally observed expression.

ANSWER TO PROBLEM: 32.10

$$N(t) = (N_0^{1/2} - 2At)^2; \quad t_{1/2} = 0.146 \frac{N_0^{1/2}}{A}.$$

This says that the half life depends on how much material you begin with, so the half life in this case, is not a constant.

TO THE OWNER OF THIS BOOK:

We hope that you have found *Calculus Problem Workbook for Hecht's Physics: Calculus* useful. So that this book can be improved in a future edition, would you take the time to complete this sheet and return it? Thank you.

School and address: _____

Department: _____

Instructor's name: _____

1. What I like most about this book is: _____

_____

_____

2. What I like least about this book is: _____

_____

_____

3. My general reaction to this book is: _____

_____

4. The name of the course in which I used this book is: _____

_____

5. Were all of the chapters of the book assigned for you to read? _____

   If not, which ones weren't? _____

6. In the space below, or on a separate sheet of paper, please write specific suggestions for improving this book and anything else you'd care to share about your experience in using the book.

_____

_____

_____

_____

_____

Optional:

Your name: _____ Date:

_____

May Brooks/Cole quote you, either in promotion for *Calculus Problem Workbook for Hecht's Physics: Calculus* or in future publishing ventures?

    Yes: _____ No: _____

    Sincerely,

    Zvonimir Hlousek
    Regina L. Neiman

FOLD HERE

BUSINESS REPLY MAIL

FIRST CLASS   PERMIT NO. 358   PACIFIC GROVE, CA

POSTAGE WILL BE PAID BY ADDRESSEE

ATT: *Zvonimir Hlousek & Regina L. Neiman*

Brooks/Cole Publishing Company
511 Forest Lodge Road
Pacific Grove, California 93950-9968

FOLD HERE

Need help solving the text's problems?
Introducing *Brooks/Cole Exerciser (BCX) v 2.0 for DOS, Windows, and Macintosh*

BCX is an easy-to-use software program that helps you solve the types of problems that appear in Hecht's *Physics: Calculus*

Features

- Each section in *BCX* begins with a problem from the text. *BCX* walks you through solving this problem.
- For each *BCX* problem, you are asked to indicate the correct solution. Choosing a correct answer prompts another problem, but choosing an incorrect answer repeats the question and offers a hint.
- A second wrong answer prompts *BCX* to provide the complete solution and an illustration of the concepts behind the problem.
- *BCX* 's record-keeping capabilities allow you to track how you are doing and to rework problems to improve your score.
- You can print your score or view it on the screen.
- The "Bookmark" feature allows you to quit the program halfway through a section and to return easily to that exact point.

Price: $20.25
DOS: ISBN: 0-534-33987-5. **Mac**: ISBN: 0-534-34171-3. **Windows**: ISBN: 0-534-34168-3.

System Requirements
DOS: 80x86 or later. DOS 3.2 or later. EGA, VGA, or SVGA monitor. 400 KB free RAM. 2.5 MB available on hard drive.
Macintosh: Mac SE or later, but not native PowerMac. System 6.0.5 or later. Monochrome or color monitor. 1 MB available on hard drive.
Windows: 80x86 or later. Windows 3.1 in Enhanced Mode or Windows 95. VGA or better graphics card. 1 MB free memory. 2.5 MB available on hard disk.

To order, please use the attached coupon or call 1-800-354-9706.

FOLD HERE

- -

BUSINESS REPLY MAIL

FIRST CLASS   PERMIT NO. 358   PACIFIC GROVE, CA

POSTAGE WILL BE PAID BY ADDRESSEE

ATT: MARKETING

**Brooks/Cole Publishing Company
511 Forest Lodge Road
Pacific Grove, California 93950-9968**

FOLD HERE

- -